WEBER CARBURETORS

PAT BRADEN

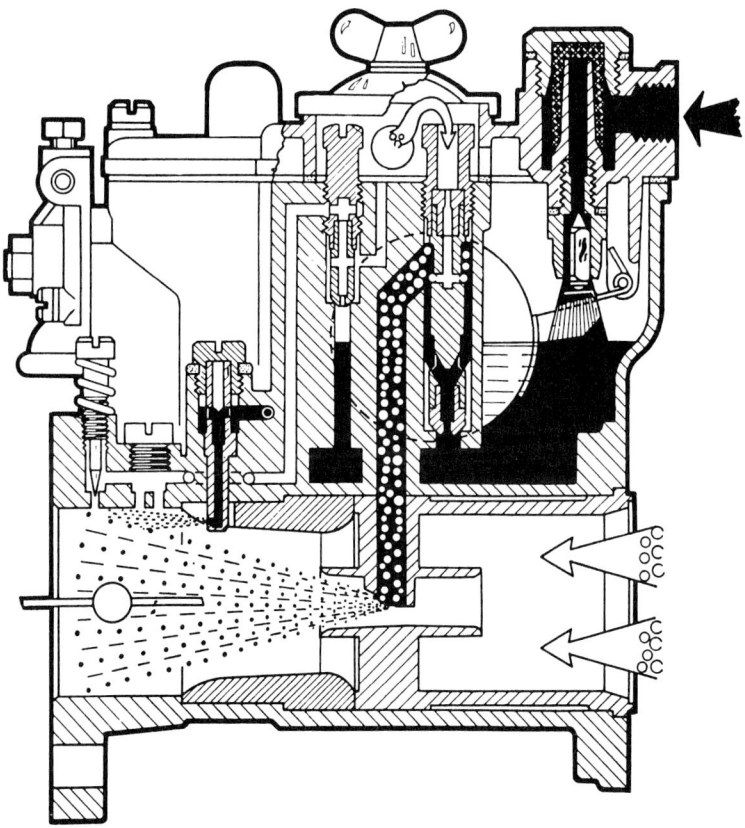

Notice: The information in this book is true and complete to the best of our knowledge. All recommendations on parts and procedures are made without any guarantees on the part of the authors or the publisher. Tampering with or altering any emission-control device is in violation of federal regulations. Authors and publisher disclaim all liability incurred in connection with the use of this information.

All rights reserved. No part of this work may be reproduced or transmitted in any form by any means, electronic or mechanical, including photocopying and recording, or by any information-storage or retrieval system, without written permission from the publisher, except in the case of brief quotations embodied in critical articles or reviews.

ACKNOWLEDGMENTS

Regardless of who takes the credit, no book is ever the work of a single author. Input for this book goes back over two decades, when owning a Weber equipped car was only a dream. Along the way, many friends—nameless here—have contributed, unknowingly, to the body of knowledge in this book. For the immediate effort, my special thanks to Don Prieto, Danny Johnson, Phil Deushene, Corky Bell, Brian Hernon, Gerry Rothman, Bruno Ratensperger and Jim Inglese. Thanks to White Eagle for the Gunson Colortune. John Shankle deserves thanks for his stewardship of Alfa all these years, and the conversion kit featured in Chapter 7. Steve Murphy and John Concialdi offered all the carburetors of Chapter 5, and informed advice. Garry Polled has my undying gratitude for infusing the manuscript with his special Weber wisdom, making it much better in the process.

My gratitude to Ron Sessions and Glen Grissom of HPBooks for their assistance and encouragement.

A book in work becomes part of the household. Kay learned not to bother daddy at the computer, and Lee was born near the manuscript's completion. For her understanding, patience and encouragement, I dedicate the book to Cheryl, my wife.

HPBooks are published by
The Berkley Publishing Group
200 Madison Avenue, New York, NY 10016
© 1988 Price Stern Sloan, Inc.
Printed in the U.S.A.

10   9   8   7   6   5   4   3

Cover photo by Bruno Ratensperger courtesy Inglese Induction Systems

Library of Congress Cataloging-in-Publication Data

Braden, Pat, 1934—
   Weber carburetors.

   Includes index.
   1. Weber carburetors.   I. Title.
TL212.B63   1988   629.2'533   87-21266
ISBN 0-89586-377-4

CONTENTS

1. **THE MAGIC OF WEBER CARBURETORS** 4
   - History 6
   - Early Racing 8
2. **WEBER CONSTRUCTION** 10
   - Family Tree 11
3. **CARBURETOR OPERATION** 14
   - How an Engine Works 15
   - Weber Design 20
   - Carburetor Circuits 22
   - Emission Control 26
4. **TROUBLESHOOTING** 30
   - Fault Identification 32
   - Main Circuit 34
   - Progression Circuit 39
   - Idle Circuit 40
   - Enrichment Circuits 41
   - Repair Methods 42
5. **REPAIR** 44
   - IMPE 45
   - DGAV 56
   - DFT 70
   - DCOE 84
   - IDA3C 100
6. **MODIFICATIONS** 115
   - Compression 116
   - Ignition Timing 117
   - Volumetric Efficiency 119
   - Manifold Selection 126
   - Linkage Geometry 133
   - Weber Selection 134
7. **WEBER CONVERSIONS** 137
   - DGAV on MGB 137
   - DCOEs on Alfa Romeo 143
   - IDFs on Small-Block Chevy 153

PRODUCTION SPECIFICATIONS 160
SUPPLIERS LIST 170
METRIC CONVERSIONS 172
INDEX 174

1
The Magic of Weber Carburetors

V12 of 166 Mille Miglia Ferrari produced 160 HP at 7200 rpm with compression ratio of 9.5:1 and three twin-choke 36 IF4/C Webers.

PERFORMANCE PEDIGREE

The classic Italian automobile of your dreams probably has Weber carburetors. And, if you're dreaming of the highest performance English cars, such as Aston Martin, Lotus and racing Jaguars, you'll find Webers on them, as well. German BMWs, Porsches and hot-rodded VWs have them. Modified big-block Chevrolets bristle with Weber air horns. No other carburetor has been so closely identified with high-performance engines as Weber. No other carburetor, in fact, has the "performance mystique" of a Weber.

Along with the mystique, there's also plenty of misinformation. Many people have heard that Webers are hard to tune, impossible to synchronize, and utterly mysterious in their workings. This misinformation can be countered; that's what this book is about.

A Weber is versatile. It can be applied as an economy carburetor or an all-out fire-breather—even though it still looks the same from the outside. That, in a nutshell, is the reason for a Weber's versatility—it's an adaptable carburetor.

Its components can be selected or changed for your specific application. But this adaptability can be two-edged. You can change so many variables that you may lose the baseline tune and driveability suffers. Again, that's where this book will be helpful. I'll show you how to select, install and tune Webers for an application with an ordered plan. You'll learn which changes to make and what their effects will be.

Mass-Produced Carburetors—Most Rochester and Holley carburetors are in this category. They have few replaceable pieces to keep manufacturing costs (and variables) low. If you've got a stock carburetor on a small-block Chevy, the engineers and designers at Rochester have pretty well set it up for you permanently, because its venturis and most of its jets are cast-in.

EQUIVALENT RATIO TABLE			
A/F (Air/Fuel)	F/A (Fuel/Air)	A/F (Air/Fuel)	F/A (Fuel/Air)
22:1	0.0455	13.1	0.0769
21:1	0.0476	12.1	0.0833
20:1	0.0500	11.1	0.0909
19:1	0.0525	10.1	0.1000
18:1	0.0556	9.1	0.1111
17:1	0.0588	8.1	0.1250
16:1	0.0625	7.1	0.1429
15:1	0.0667	6.1	0.1667
14:1	0.0714	5.1	0.2000

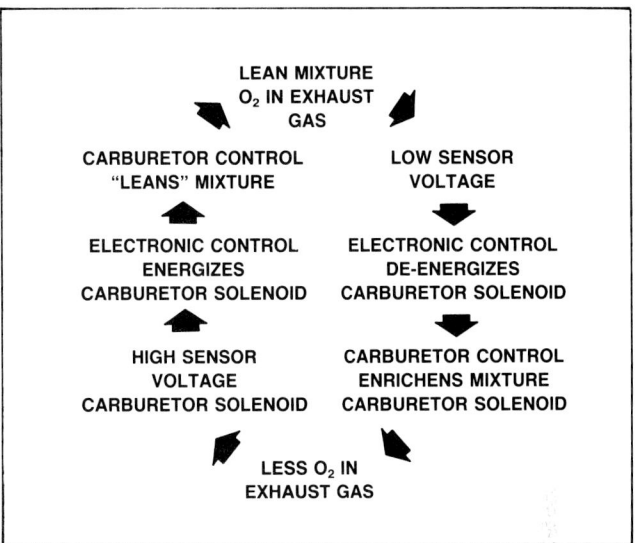

Use table to match equivalent Air/Fuel to Fuel/Air ratios.

ECU determines correct air/fuel mixture by processing oxygen-sensor voltage in closed-loop operation. Drawing courtesy Rochester Products Division, GM.

EQUIVALENT RATIOS

The ratio that expresses the amount of fuel to the amount of air flowing together into an engine is called the *fuel/air (F/A) ratio*. This specifies the amount of mixture in pounds of fuel divided by pounds of air. Engines use much more air than fuel, so fuel/air ratios are small numbers such as 0.068. This equals 1 (one) pound of fuel divided by 14.7 pounds of air.

Some people are more familiar with stating mixture combinations as an *air/fuel (A/F) ratio*. This is just another expression to state the same value. In our example, this means the engine uses 14.7 pounds of air for each pound of fuel (14.7:1).

Take the reciprocal of the F/A ratio to get the equivalent A/F ratio. That is, divide 1 by the F/A. So, dividing 1 by 0.068 F/A ratio equals a 14.7:1 A/F ratio. Or look at the nearby Equivalent Ratio Table to save wear and tear on your calculator or brain cells.

Mass-produced carburetors are designed to work in the real world of strict emission controls and must deliver superior fuel economy. They're usually not the best carburetors for performance applications. Internal combustion engines burning gasoline run most efficiently at an air/fuel mixture that is roughly 14.7 parts air to one part fuel by weight (14.7:1 air/fuel ratio). This ideal mixture ratio is called *stoichiometric*. It's the mixture ratio at which all gasoline is burned using all the available air.

Most mass-produced carburetors produced since 1981 employ *closed-loop control* to keep the mixture close to stoichiometric. This mixture control method is discussed more thoroughly in HPBooks' *Rochester Carburetors* and *Holley Carburetors & Manifolds*.

Very briefly, these *feedback* carburetors use a system of engine sensors—oxygen content in the exhaust stream, engine temperature and others—to supply data to a microcomputer. Usually called the *Electronic Control Unit* (ECU), it adjusts the fuel/air mixture in the carburetor using a *solenoid-controlled valve* or a *duty-cycle solenoid*.

But, if you've punched the engine out a few cubic inches, increased its stroke or changed the camshaft(s), you'll need to modify the carburetor to handle the increased airflow that results. One option is to buy a stock carburetor that flows more air. On those carbs, the internals are set up to run a stochiometric mixture for an "average" engine.

To change the profile of fuel/air ratios, for example, so the mixture will be more rich from 1500—2500 rpm, leaner at mid-range, and excessively rich over 7000 rpm, almost requires re-engineering the carburetor.

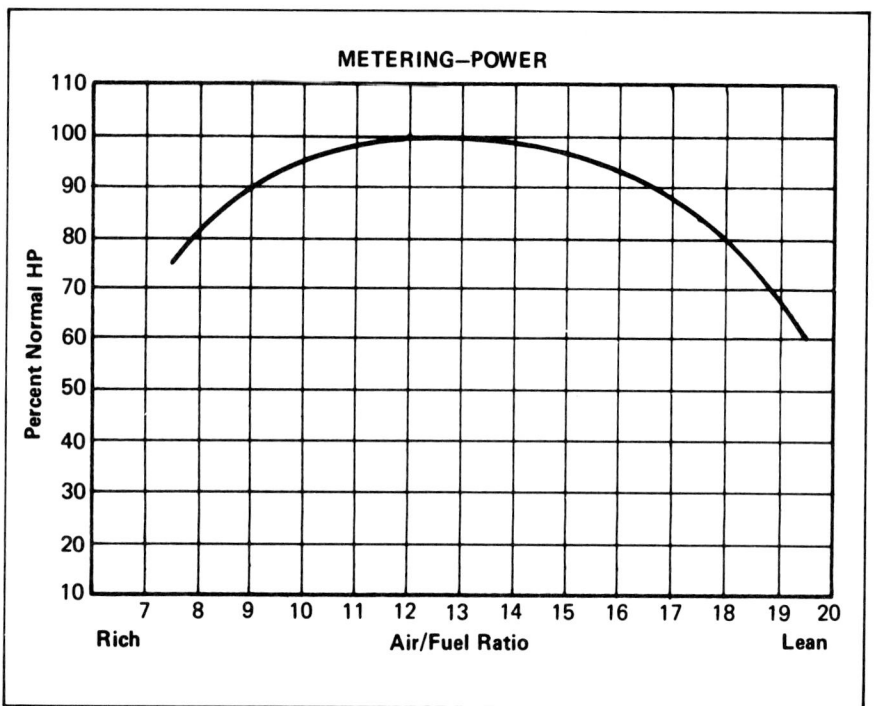

Power curve of typical engine relative to air/fuel ratio: Curve is flat between 11:1 and 14:1, meaning engines produces good HP over broad mixture range.

Now, I don't mean to sell all mass-produced carburetors short. Some can be tuned for levels of increased performance. But most Webers offer the tuning advantage (remember its double edge) of being *easily* adapted to whatever performance application you may have.

WEBER VARIETY

Webers are a veritable candy store when it comes to variety. On many Webers you get to choose different factory-calibrated values for:

- Main, idle, intermediate and accelerator-pump jets.
- Air correction and emulsion tubes for the main and intermediate circuits.

Some even let you change the size of the throat—*venturi*—itself. Furthermore, the new parts will have been as precisely calibrated as the original piece you've removed.

But we don't have to limit our choices just to pieces on a specific carburetor. Webers offer even more versatility than that. Across the Weber product line, you can also choose whether you want:

- The float bowl between or beside the throats.
- Electric- or water-heated automatic or manual chokes (or no choke at all).
- The two venturis in-line with or transverse to the engine.
- The second throttle to open simultaneously or progressively with the first.

Because you can end up reconfiguring just about every part of the carburetor installation, Webers force you to know what you're doing. When you increase venturi size, you must change the main jet, air-correction jet and emulsion tube sizes to supply the correct fuel flow. And, if you change the main jet, you may also have to change the idle and intermediate jets.

If you select simultaneous-opening throttles, you must pay special attention to the intermediate circuits to retain driveability. And, if you select a Weber with a different float-bowl location, be sure fuel sloshing difficulties don't increase with the new placement.

Moreover, you can't have just a vague idea of what steps and adjustments to take when tuning. The Weber carburetor is a precision instrument, capable of delivering fuel to the engine with superb accuracy. Your knowledge of carburetion has to be as precise if you're to take full advantage of what a Weber can do. This book will contribute to your knowledge so you can efficiently select, install and tune Webers.

Clearly, the likes of Ferrari and the Maserati brothers know what they're doing when it comes to carburetion: That's one reason why classic Ferraris and Maseratis have Webers. And the catalog of exotic Weber equipped cars clearly adds to the Weber performance image. But it's not just mystique or image: Webers deliver. It's clear, from the little historic information available, that Weber carburetors have a long-term reputation that they work better than their competition. Here's a quick review.

WEBER HISTORY

Origins—Born in Turin, Italy in 1889, Edoardo Weber worked in the Fiat plant briefly before taking the job of service manager for a Fiat dealership in Bologna, in 1914. In 1920, gasoline was very expensive in Italy, so he went into business for himself making kerosene-conversion kits for trucks. In 1925, he switched to producing hop-up kits for Fiats.

The earliest Weber carburetor known was part of such a kit. It was a sidedraft, double-throat carburetor that bolted to a Weber designed overhead-valve/supercharger conversion for the 501 Fiat. The carburetor throats were of different diameters. The idea of using two different-sized throats (a small one for low speeds and a larger one for high speeds) appears to have been a Weber original.

Weber used his two-throat carburetor to control whether or not the engine was supercharged. At anything less than wide-open throttle (WOT), a cut-out valve leaked the pressurized air from the supercharger to the atmosphere. The engine then ran, unpressurized, from the smaller throat. At full throttle, the cut-

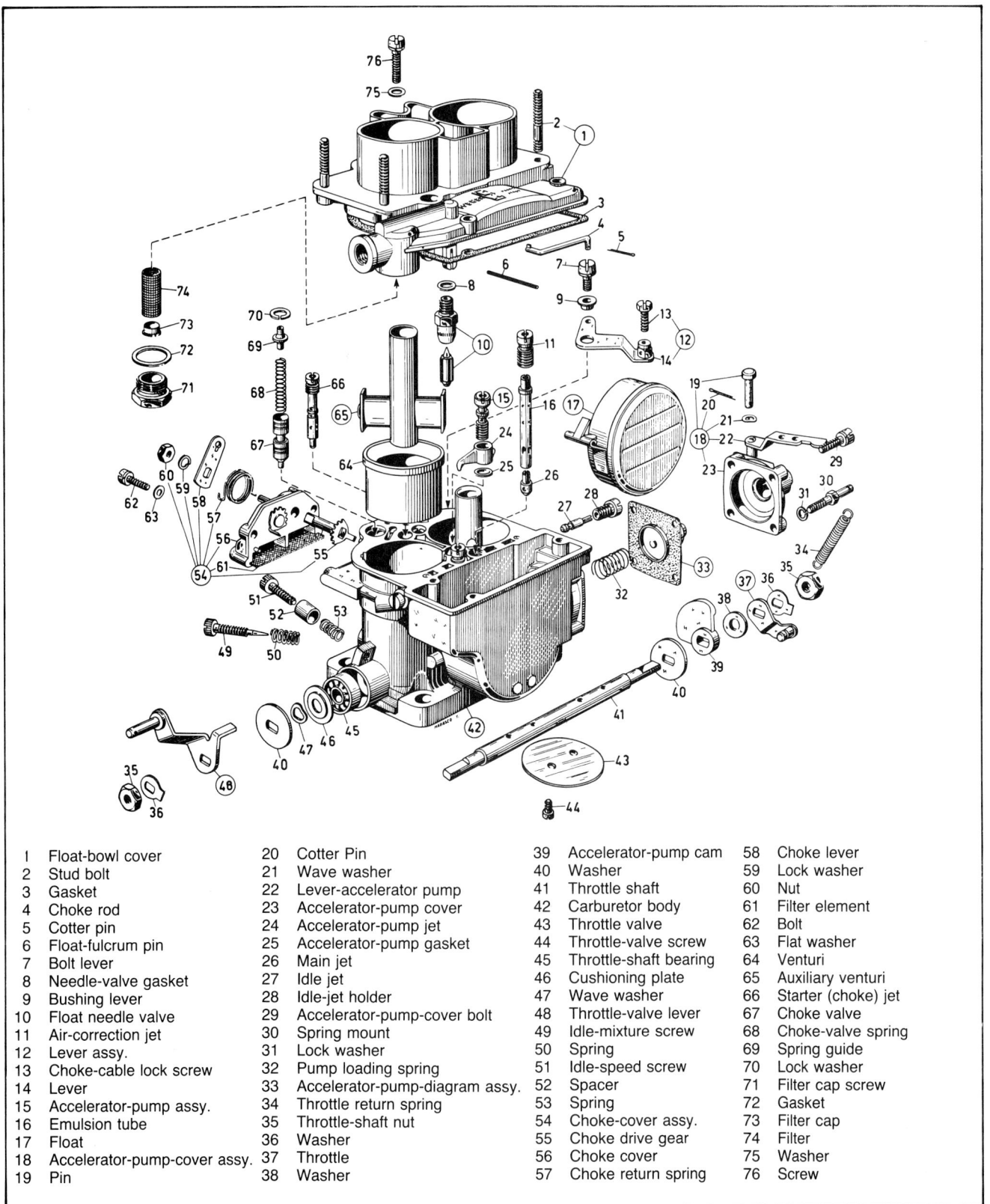

1	Float-bowl cover	20	Cotter Pin	39	Accelerator-pump cam	58	Choke lever
2	Stud bolt	21	Wave washer	40	Washer	59	Lock washer
3	Gasket	22	Lever-accelerator pump	41	Throttle shaft	60	Nut
4	Choke rod	23	Accelerator-pump cover	42	Carburetor body	61	Filter element
5	Cotter pin	24	Accelerator-pump jet	43	Throttle valve	62	Bolt
6	Float-fulcrum pin	25	Accelerator-pump gasket	44	Throttle-valve screw	63	Flat washer
7	Bolt lever	26	Main jet	45	Throttle-shaft bearing	64	Venturi
8	Needle-valve gasket	27	Idle jet	46	Cushioning plate	65	Auxiliary venturi
9	Bushing lever	28	Idle-jet holder	47	Wave washer	66	Starter (choke) jet
10	Float needle valve	29	Accelerator-pump-cover bolt	48	Throttle-valve lever	67	Choke valve
11	Air-correction jet	30	Spring mount	49	Idle-mixture screw	68	Choke-valve spring
12	Lever assy.	31	Lock washer	50	Spring	69	Spring guide
13	Choke-cable lock screw	32	Pump loading spring	51	Idle-speed screw	70	Lock washer
14	Lever	33	Accelerator-pump-diagram assy.	52	Spacer	71	Filter cap screw
15	Accelerator-pump assy.	34	Throttle return spring	53	Spring	72	Gasket
16	Emulsion tube	35	Throttle-shaft nut	54	Choke-cover assy.	73	Filter cap
17	Float	36	Washer	55	Choke drive gear	74	Filter
18	Accelerator-pump-cover assy.	37	Throttle	56	Choke cover	75	Washer
19	Pin	38	Washer	57	Choke return spring	76	Screw

Fiat Dino Weber 40 DCNF 3 has large variety of interchangeable parts, including venturi (64) and auxiliary venturi (65), plus all jets that control air and fuel flow.

Weber replacement of original carburetors can match original installations. Here, JAM Engineering's twin 32/36 DGEV's have replaced the stock Zenith INATS on a Mercedes. (Photo Courtesy of JAM Engineering.)

Triple-throat Weber fitted to Alfa Romeo Type 159 Grand Prix car of 1951: 1.5-liter engine used two-stage supercharging to develop 404 HP at 11,000 rpm. This was enough punch to make Alfa World Champion in 1951. Photo courtesy Alfa Romeo Archives.

out valve closed and the engine was supercharged through the second throat of the carburetor.

Weber discovered that his design of two different-sized throats could be used without a supercharger to gain fuel economy. The larger throat would be used for maximum performance and the smaller for economy. It was not long before two-throat Weber carburetors became common conversions for Fiat 501s used as taxicabs in Italy in the late '20s.

Early Racing Efforts—Weber was certainly a pioneer of two-throat carburetion, and he began using two throats of identical sizes to feed separate cylinders of the engine. The 1931 Maserati 1100cc Grand Prix car featured the first two-throat sidedraft Webers, the 50 DCO.

The earliest Webers you're likely to find date from the classic Alfa Romeo cars of the same era. These Alfas were straight-8s with twin camshafts and either one or two superchargers, depending on displacement. The most successful of these Alfas, the Tipo B 2900 of 1933, carried a pair of Weber single-throat updraft carburetors sitting beneath its two superchargers.

Another early Weber was the three-throat 50 DR3C fitted to the Type 158 Alfa, a Grand Prix car that would ultimately (in 1951) use two-stage supercharging to develop 404 HP from just 91.5 CID (cubic inch displacement). Three throats were necessary to supply even air/fuel flow along the entire length of the supercharger vanes.

The success of Weber equipped cars made Weber successful as a carburetor manufacturer. His early experimental association with Fiat, Maserati and Alfa Romeo finally attracted attention worldwide. In addition, Weber wisely courted the Italian car manufacturers to become their supplier of carburetors. He was so successful that virtually every Italian car, from the smallest Fiat to the most exotic Ferrari, carries Weber carburetion.

In addition to good design and salesmanship, Weber offered fabulous customer support. According to Alf Francis, in his book, *Alf Francis—Racing Mechanic,* Weber technology came to England in 1952 when Alf asked to have a pair of Webers fitted to an Alta. An immediate 7-HP increase was recorded on the dyno. After some fiddling (and the design and fabrication of an entirely new intake manifold), the 117-HP Alta was pumping out 143 HP. The total bill to Francis was only for the two carburetors, no charge for days of dyno testing or for fabricating the intake manifold.

Finally, Weber has consistently maintained excellent quality control of its components. For example, each jet is individually flowed, and the flow results determine the number marked on the jet.

After Edoardo Weber's death in 1945, Fiat began taking an active interest in the welfare of the company and succeeded in buying more than half of it by 1952. Fiat was very active in racing during this period, first with Siata and then Abarth. Both those marques used Weber carburetion almost exclusively, and gained plenty of headlines for themselves and Weber as well.

U.S. Introduction—In the U.S., Webers were first imported by the Geon company during the early '50s. They were virtually mandatory accessories for drivers of Jaguars and MGs hoping to win the public-road races that characterized the competition of the period.

It wasn't long, however, before stateside owners of Alfa Romeo Giulietta and Opel 1.9-liter cars discovered that the Fiat DCD Weber was a bolt-on fit.

Look familiar? BRE Datsun 510 used "work-alike" of Weber—pair of 44mm PHH Mikuni/Solexes. Car was formidable competitor.

And, it supplied more power and better driveability than the original-equipment Solex. Further, the DCD's replacements, the DGV, and DFAV, are bolt-on replacements for the Holley 5200 carburetor series. So, Webers began appearing on Ford products as replacement carburetors for this Holley.

Also contributing to Weber's success is the large variety of carburetor types the company offers. Carburetor styles currently include single-, double- and triple-throat downdraft and dual-throat sidedraft units. (The single-throat sidedraft is a collector's item.) The dual downdraft carburetors have either simultaneous or progressive throttle linkages and float bowls placed either to the side or between the throats. As a result, you can create an almost ideal carburetor setup with Webers, no matter how odd the application.

And, in many applications, you'll be able to change all of the most important features of the carburetor without removing it from the engine. DCOE sidedraft Webers are especially notable for the accessibility of their jets. On most, the jet carriers are accessible even with the air cleaner attached.

Weber design principles have been adopted by several other manufacturers. Weber work-alikes are produced in Italy (Dellorto) and Japan (Mikuni).

As Weber's applications in the U.S. have grown, so has its distribution channels. Geon, an early national distributor, became BAP/GEON. But that firm has now been replaced by INTERCO as the distributor for the states east of the Mississippi. Redline, Inc. is the official Weber importer for the U.S. Several independent importers, such as TWM Induction in Goleta, CA, simply buy their carburetors directly from Italy. An entire industry of Weber specialists has grown up, and most major U.S. cities have at least one shop specializing in Webers.

Over the years, Weber has grown from a small company in Italy to the premier worldwide supplier of specialist carburetors. A standard of quality continues to distinguish Weber products from their competition. Edoardo Weber wouldn't have it any other way.

2
Weber Construction

Webers make boats go faster, too. Forest of velocity stacks conceals twin big-block Chevys topped with 48 IDA Webers. Photo courtesy Inglese Induction Systems.

WEBER CONSTRUCTION

Before going into their details and differences, let's look at what many Webers have in common.

Imagine yourself during practice on a road-race course, needing to extract a one-second better lap time to assure a win. You're sure a slightly larger main venturi will give your DCOE-equipped engine that elusive extra power to bring you home a winner.

But fitting a larger venturi also means you may have to change the main jet, main air-correction jet and, perhaps, even the emulsion tube. Otherwise, the car could run lean and you'd risk melting some pistons. And, since you have a moment, you want to increase the punch out of tight corners by upping the fuel output of the accelerator pump.

If you could also enrich the progression circuit by fitting a slightly larger jet and smaller air-correction jet, you'd have just a little bit better transition throttle control in the esses. But, you don't have time to remove the carburetor. With a DCOE Weber, you can make these adjustments without removing the carburetor from the engine.

Here are some details typical of most Webers used for high performance:

• Major jets are accessible from the outside of the carburetor.

• Removable components are accessible with the float bowl cover removed.

• Boost venturi and main venturi are removable.

• Main jet is carried in a removable holder, instead of being screwed directly into the body of the carburetor.

• Air-correction jet and emulsion tube are part of a single assembly, or are

assembled one on top of the other.
- Separate jets control fuel flow to the idle, main, accelerator-pump and starting circuits.
- Starting circuit is essentially a separate carburetor with its own venturi and main jet.
- Accelerator pump is a piston acting in a cylinder.
- Jet thread diameters are unique so the jets cannot be replaced improperly in the carburetor.

Now, these characteristics are not typical of every Weber. In fact, some Webers have cast-in venturis, diaphragm-type accelerator pumps and butterfly valves for starting the engine. These fall into the *production types* of Webers. But, in general, the design traits above apply to the most popular *high-performance* Webers.

WEBER FAMILY TREE

It's not easy to organize Webers into neat families for two reasons. In the first place, Weber didn't set out to construct a tight little family of carburetors. He was interested in building carburetors that best matched the current need. As a result, there never was an overall scheme of things. Second, he named carburetors in his native Italian. So if you're unfamiliar with words like *doppio, corpo* or *orizzontale,* Weber terminology can be confusing.

To further muddle the issue, Weber typically created different model numbers whenever he could. He would give carburetors unique names even though they differed only in jetting or some similarly undistinguishable feature. Thus, a DCN fitted to the Lamborghini was a DCNL. If fitted to a Ferrari, the same carburetor became a DCNF! Another twist is represented by the IDA carburetor. It may have either two or three throats, yet the same identification letter.

While it would be helpful to have a neat chart of Weber terminology, it's simply impossible. Nevertheless, as chaotic as the terminology is, there are some common threads in how Weber named his models.

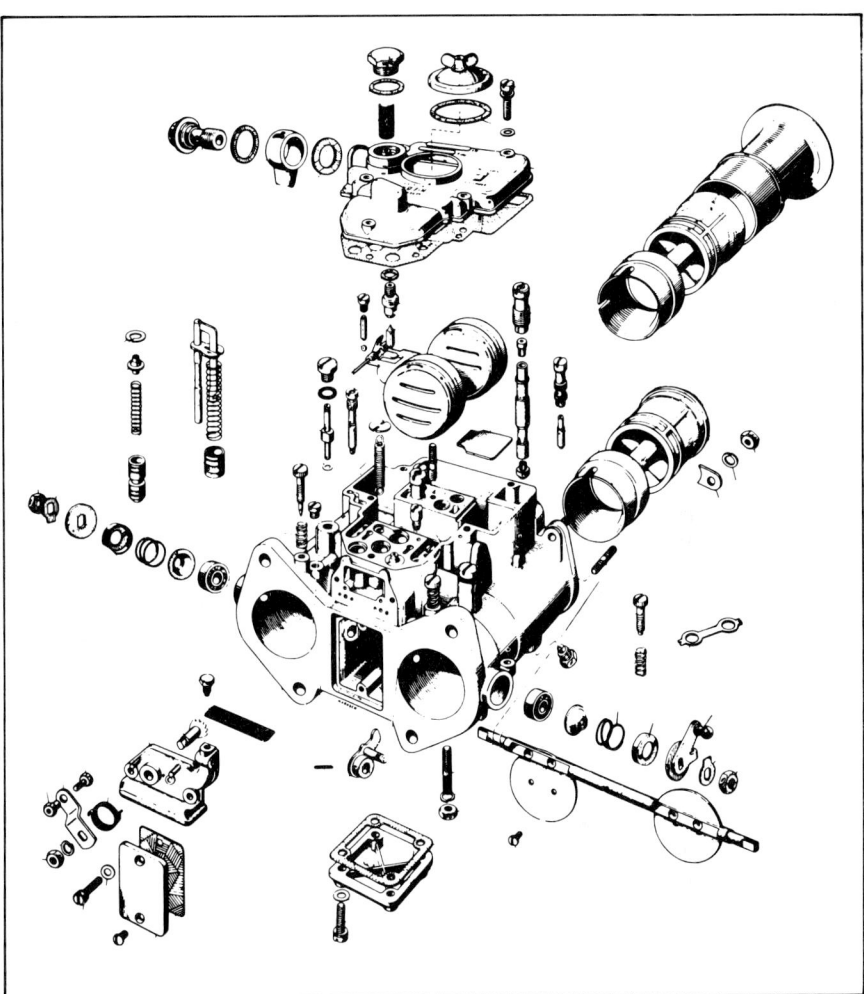

DCOE is model most think of when referring to Weber carburetors. Sidedraft design is perfect setup for in-line engines.

Model Numbers—A Weber carburetor is designated with a series of numbers, then letters, then, sometimes, more numbers. The first set of numbers is called a *prefix,* the second set, a *suffix.* The letters in between seem to be the first letters of Italian words, but not always.

The prefix always indicates the size of the throttle plate. For example, the 40 DCNF and 4O DCOE both have 40mm throttle plates. The rule that the prefix refers to throttle bore diameter is only one of two sure-fire rules in Weber nomenclature. The other is the jet-numbering convention described later.

The *letters* immediately following the prefix refer, in Italian, to the general type of the carburetor. Only a few terms are consistent: DC means *doppio corpo,* or double-throat; V means *verticale* and O means *orizzontale* (vertical and horizontal, respectively). The I in IDA seems to mean *invertito,* or inverted. Yet, there are no up-draft, or inverted Webers: the IDA is a *downdraft.* Other even less-consistent single-letter designations are:

Porsche 904 sported Webers on Type 547 engine. Photo by Alex Gabbard.

- **E** Die-cast carburetor
- **F** Ford (or Ferrari?) application
- **V** Carburetor with a power valve
- **A** Water-operated automatic choke

Any *numbers* following the letters are variations of the basic type: the 45 DCOE9 is a variation of the 45 DCOE. But, note that, to date, there are 10 known variations of the 45 DCOE9!

So Weber carburetor nomenclature is chaos. To return to the IDA, there may be three, as well as two throats, and there is no choke at all.

Here is one last example of how the terminology collapses. The 40 DCO carburetor fitted to the early Alfa Romeo Giulietta Veloce follows the nomenclature exactly. It is a double-throat horizontal carburetor with 40mm throttle bores. The 40 DCOE of later Alfa use, however, is neither die-cast nor equipped with an electrically operated choke. In this case, the E indicates a trapezoid mounting bolt pattern instead of the rectangular one of the DCO.

Jet Size Designation—When it comes to jet sizes, Weber is refreshingly consistent in numbering the replaceable jets. Air-correction jets, main jets and idle jets all have numbers expressing their actual flow rates in hundreths of a millimeter (0.01mm). A 115 main jet flows the same amount as a perfect hole 1.15mm in diameter.

Emulsion Tube Designation—If you're sizing emulsion tubes, however, there is absolutely no hope in linking function and nomenclature. Numbers for emulsion tubes are assigned chronologically. You can be sure an F-45 emulsion tube was designed after the F-44, but that's about all.

Some **idle jets** have F-numbers that indicate their ability to emulsify fuel. There has been some deciphering to make order of their numbering. According to Phil Deuchene of BAP/GEON, the following chart applies:

Idle-Circuit Emulsion Calibration										
Rich (small)										Lean (large)
F6	F12	F9	F8	F11	F13	F2	F4	F5	F7	F1	F3

Austin-Healey 3000 with 100-6 engine in works rally trim sported three sidedraft Webers. Photo by Alex Gabbard.

Carburetor Types—The accompanying chart puts some order to the main carburetor types. It's not an official Weber organization, but is developed from study of the carburetors themselves. It groups all Weber carburetors into two major types—vertical and horizontal—and then subdivides each major division according to the number of throats.

Within carburetors of two throats, there is a further subdivision, depending on whether or not one or two throttle shafts are used; the two-shaft layout permits progressive secondary throttle opening (the popularity of this design explains why there are so many different types in the category), while the one-shaft, two-throat carburetor has simultaneous throttle openings.

Of course, this is not the only way you can organize Webers. While there are some significant exceptions, Webers are divided generally between *sport* and *production* carburetors. Obviously, the IDA and its work-alikes, with no starting circuit and rather large throats, are made for passing such large volumes of air that you can just about assume racing tune. Similarly, the DCO-series carburetors have been used on so many outright racers, they have earned a place in the sporting category, even though they are also used on regular production cars.

No doubt there are several other ways to group Webers, but such groupings probably have very limited application. For the purpose of figuring out which Weber would work best on a specific engine, the first chart is most helpful.

There's a final item to note about Weber carburetor construction. They are veritable road maps when it comes to figuring out how the fuel flows. Weber's casting techniques are fine enough that they are able to cast fuel passages in deep relief. As a result, you can follow most fuel flow paths just by looking.

It's very instructive, before disassembling a Weber, to trace the fuel paths simply by following the cast-in passages. As the carburetor comes apart, you'll discover the remainder of the paths and uncover much of how a Weber works in the process.

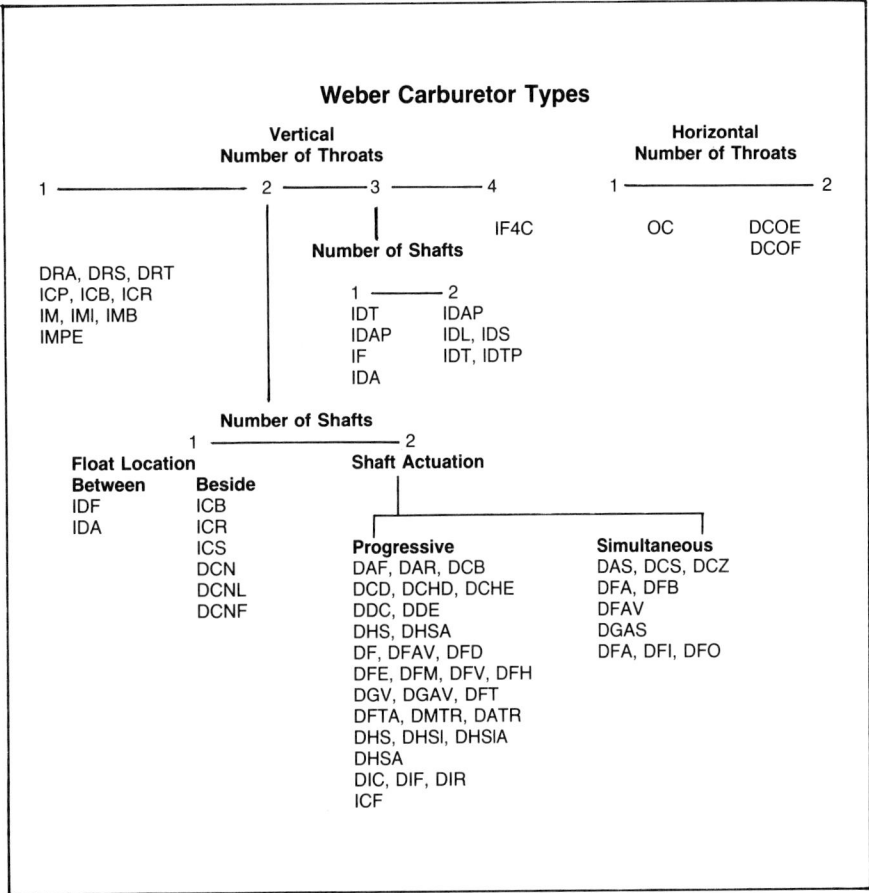

If you sort between sport and production applications, the carburetors shake out like this:

SPORT			PRODUCTION		
IDT	IDAP	IF4C	DRA, DRS, DRT	DAF, DAR, DCB	DAS, DCS, DCZ
IDAP	IDL,	IDS	ICP, ICB, ICR	DCD, DCHD, DCHE	DFA, DFB
IF	IDT,	IDTP	IM, IMI, IMB	DDC, DDE	DFAV
IDA			IMPE, ICB	DHS, DHSA	DGAS
			ICR, ICS	DF, DFAV, DFD	DFA, DFI, DFO
DCOE	DCOF	DCOE		DFE, DFM, DFV, DFH	
				DGV, DGAV, DFT	
IDF		DCN		DFTA, DMTR, DATR	
IDA		DCNL		DHS, DHSI, DHSIA	
		DCNF		DHSA	
				DIC, DIF, DIR	
				ICF	

Note: Groupings of the first chart have been retained to make comparison easier.

3
Carburetor Operation

Know your application before selecting Webers. This '32 Ford sports four chrome 48 IDAs. Efficiency is the goal, and setup looks great.

This chapter explains some of the variables affecting the air and fuel requirements of an internal combustion engine. For example, it's helpful to understand what happens in the intake passage of an engine in order to change anything for the better, which is what we plan to do by fitting a Weber. The task is to become aware of these requirements so you can select the correct carburetor.

Now, a carburetor is not a stand-alone unit. To understand how to select and tune a Weber successfully, you need to understand how a carburetor affects engine operation.

First, there are other sources that examine the dynamics of the top end of an engine in more detail than I can here. A standard work is *The Scientific Design of Exhaust and Intake Systems* by Philip H. Smith (London, 1963, G.T. Foulis & Co. Ltd). It's available from Classic Motorbooks, P.O. Box One, Osceola, WI 54020. If you want to immerse yourself in tuning, that's one place to start. For some, reading Smith's book will seem like learning how to make a watch when all you wanted to know was the time.

Over the years, there has been a significant change in the complexity of an engine. Early engineers were most concerned with using the laws of nature just to get an engine to work. They didn't have any pat answer to the best way of turning the potential energy in gasoline into mechanical energy. At first, these early designers were more concerned with just getting fuel and oxygen into an engine and igniting it.

Today's engineers can explain the complexities of a modern engine in exacting detail. But when reviewing these complexities to get an overall un-

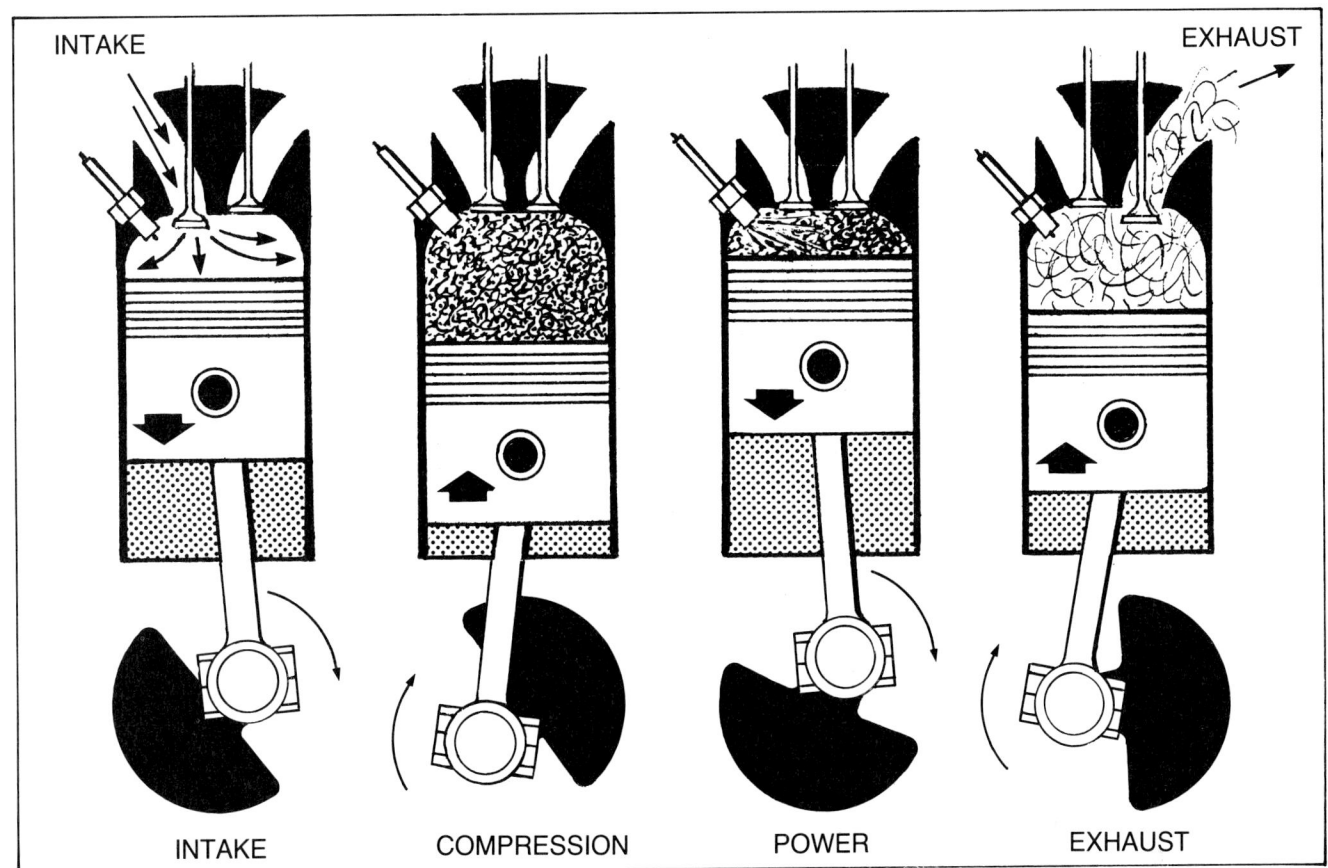

On INTAKE stroke, intake valve opens (left) and piston moves down to draw in combustible mixture. COMPRESSION stroke: both valves closed, piston moves up, compressing combustible air/fuel mixture. Sparkplug ignites combustible mixture; expanding gasses drive piston down on POWER stroke. Burned gases are pushed out by rising piston past open exhaust valve (right) during EXHAUST stroke.

derstanding of how an engine works, we risk oversimplifying. A modern engine takes concentrated study to know it really well.

So, what follows is not an exhaustive study of engine design and how it affects carburetion. It is, instead, a survey of those essentials of engine operation that influence selecting and tuning Webers.

HOW AN ENGINE WORKS

The basic idea of the conventional internal combustion engine is to heat some gas—an *air/fuel mixture*—so it expands and is converted to mechanical work by pushing against a piston. The moving piston connects to a mechanical linkage (connecting rod and crankshaft) that transmits this up-and-down motion—*reciprocating motion*—into turning, or *rotary,* motion. (A change in motion isn't necessary in a rotary engine such as the one used in a Mazda RX-7.)

The essentials, which are what the early engineers chased, are to mix atomized gasoline with air, trap the mixture in a cylinder, squeeze it a bit and ignite it. The burning air/fuel mixture expands and then drives the piston down the cylinder.

One of the physical laws of nature is that heated gases expand. The pressure developed by the expanding mixture is the source of an engine's power. This pressure is measured in pounds per square inch (psi), and ranges over 3000 psi for a modern engine. The measure is called either *mean effective pressure,* or *brake mean effective pressure (BMEP)*.

You can increase BMEP in several ways. One is by increasing the actual *compression ratio* of the engine. It is a measure of how much the air/fuel mixture is compressed in the cylinder. Compression ratio is derived by dividing the volume in a cylinder with the piston at the bottom of its stroke (bottom dead center or BDC) by the volume with the piston at the top of its stroke (top dead center or TDC). *Clearance volume* is the name of the volume above the piston at TDC.

Another method to increase BMEP is

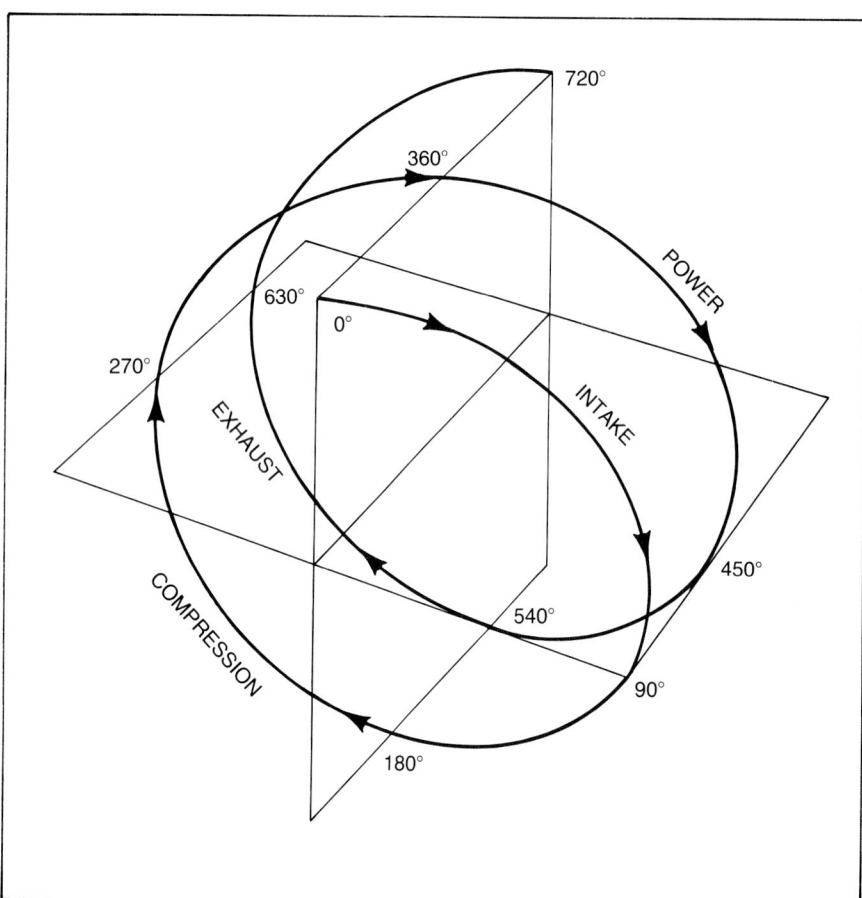

Two complete crankshaft revolutions, or 720°, are required to complete Otto cycle.

by packing more combustible mixture into the cylinder. This is examined in more detail later. But first, let's review some of the fundamentals of engine operation and then see how a carburetor contributes.

Otto Cycle—The typical automotive internal combustion engine operates in a *four-stroke cycle*. This is also called the *Otto cycle* after its developer, Dr. Nikolaus Otto of Germany. The engine requires four strokes, two up and two down—or two complete revolutions of the crankshaft, 720°—to complete one cycle. These strokes are in order: *intake, compression, power,* and *exhaust.*

To simplify the basic descriptions of these strokes, I've made one assumption: the valves are considered to open and close at TDC and BDC. In actual engines, they aren't timed to open at these points. Their timing (closing and opening) *overlaps,* see page 19.

Intake—During this stroke, the intake valve has opened (the exhaust valve is closed), the piston is moving down, and air/fuel mixture is moving into the cylinder. The downward motion of the piston creates a *partial vacuum* in the cylinder. The atmospheric pressure—14.7 psi at sea level—of air outside the engine is greater than the pressure inside the cylinder. So, the *pressure difference* between the atmosphere's pressure and that in the cylinder forces air into the engine, moving it through the carburetor (picking up gasoline vapor), through the intake manifold, past the open intake valve and into the cylinder.

Compression—Now both valves are closed and the mixture is captured in the cylinder. The piston moves up in the cylinder to TDC and compresses the mixture. The mixture becomes hotter when compressed and is ready to be ignited. At this point, the piston has completed two strokes and the crankshaft has revolved one complete turn or 360°.

Power—As the piston approaches TDC on the compression stroke, the air/fuel mixture is ignited by a spark at the sparkplug. *Combustion* occurs. As the burning air/fuel mixture tries to expand in the sealed cylinder, it drives the piston down toward BDC in the power stroke. The piston does its work during this stroke—it delivers energy to the crankshaft and causes it to turn.

Exhaust—As the crank passes BDC, the piston moves upward, the exhaust valve has opened and the burned mixture products are forced out of the cylinder. When the piston reaches TDC, the exhaust valve closes and the intake valve opens. A fresh mixture charge begins to pass the open intake valve and the Otto cycle begins again.

Early Engines & Carburetors—The earliest engines used air pressure alone to open the intake valve, which was held shut by a weak spring. As the piston moved down the cylinder, outside air pressure overcame the force of the spring and the intake valve was opened. When the cylinder was almost filled with mixture, the pressure in the cylinder approached atmospheric pressure and the weak spring closed the valve. The intake valve was then held closed by the rising pressures inside the cylinder as the piston began its compression stroke. That invention was called the *automatic intake valve.*

Early carburetors were equally simple. One early type was called a *surface carburetor,* which bubbled air through a bowl of gasoline. As the air bubbled through the gasoline, it picked up enough vapors to become an ignitable mixture.

Mixture Flow & Speed—These early engines did just cough to life. Their top speeds ranged around 1000 revolutions per minute (rpm). That must have

INTAKE

Higher atmospheric pressure helps fill low-pressure volume created by piston descending in cylinder.

AIR IS A "HEAVY" SUBJECT

Air has weight. Imagine a one square inch column of air that reaches to the top of the atmosphere for its height. That column weighs 14.7 pounds (or *1.0 atmosphere*) at sea level.

The reason you don't notice the weight is that air exerts pressure equally on everything we deal with, so it cancels out any evidence of its own weight. This is similar to how the weight of a bucket of water is insignificant when we're scuba diving. One time the weight of air is significant is when there's a *vacuum* (less pressure), such as found in an engine's intake manifold.

If we think in terms of air having weight, then we can throw out the idea that a vacuum draws in air. Instead, a vacuum is just the absence of air pressure. A vacuum isn't sucking anything; the air is pushing in on it because of its weight.

As a piston travels down a cylinder, increasing the volume of the cylinder, the weight of the outside air causes the cylinder to be filled with more air by *displacing* this volume. The same action happens when we breathe: Our diaphragm muscle enlarges the cavity around our lungs, and air pressure pushes fresh air into the lungs. Engines and humans both breathe because air has weight.

When we think of air as having weight, it's easier to explain why it acts the way it does. We know what to expect from a heavy object like a bowling ball: Once started rolling, it tries to keep rolling in a straight line, even when it hits a pin. Air acts the same way, but because it weighs less than a bowling ball, its similarities to bowling-ball behavior are much more subtle. That is, it starts, stops and changes direction more quickly. But, like the bowling ball, it offers resistance to changes in speed and direction.

---

seemed pretty fast to those who had never seen anything faster, as the first engineers had not. A thousand rpm is near 16 times a second, and that's blinding fast for a lot of iron to be jerking around.

But the search for horsepower is a search for faster engine speed. That's because horsepower is a calculated value that includes a time: the twisting force of 33,000 pounds on the end of a one-foot lever for one minute. In more tidy terms, one horsepower is 33,000 pound-feet (lb-ft) of torque per minute—not that you need to know how to calculate horsepower: The point is that the formula for horsepower includes a *value for time*. Therefore, if the same amount of twisting force is developed in less time, horsepower increases.

TERMS

Gas—Our penchant for calling gasoline *gas* confuses the issue when discussing a mixture that includes both air, which is a gas, and gas, as in *gasoline*. I use the term *mixture* throughout to refer to the combination of air and gasoline burned in an engine.

Mass—When the astronauts landed on the Moon, they could jump around like rabbits because they weighed only a sixth of what they weighed on Earth. Nothing about the astronauts themselves had changed. What changed was the force of gravity.

When we say *weight*, we're really talking about the force on our *mass* exerted by gravity. We can go to the Moon, or Jupiter and our mass won't change, though our weight will.

It's difficult to separate the idea of mass and weight because we rarely encounter a situation where the two don't equate. I've noted that they do separate company in outer space, and maybe our grandchildren astronauts won't have any difficulty with the distinction at all. But for us who have lived our lives Earthbound, the distinction between mass and gravity is subtle.

When I talk about the flow of the mixture into an engine, you need to refer to its mass, not weight. Remember that the tendency for a bowling ball to keep rolling in a straight line is the same whether we're on Earth, where the ball weighs several pounds, or on the Moon, where it would weigh several ounces. That's because its momentum (see below) is *mass-dependent*, not *weight-dependent*. Just to keep the discussion accurate, therefore, I use *mass* when talking about the forces that affect the way mixture moves.

Inertia & Momentum—Inertia can be broadly defined as the general tendency for an object to stay at rest when it's already at rest, and the tendency for it to stay in motion when already in motion. *Momentum* is defined as the product of an object's mass (m) and its velocity (v): momentum = (m)(v). Momentum changes whenever mass or velocity changes. For our purposes, if the mass or velocity of a mixture changes, then its momentum will change. Notice that I said *velocity*, not speed. Velocity is speed *with direction*.

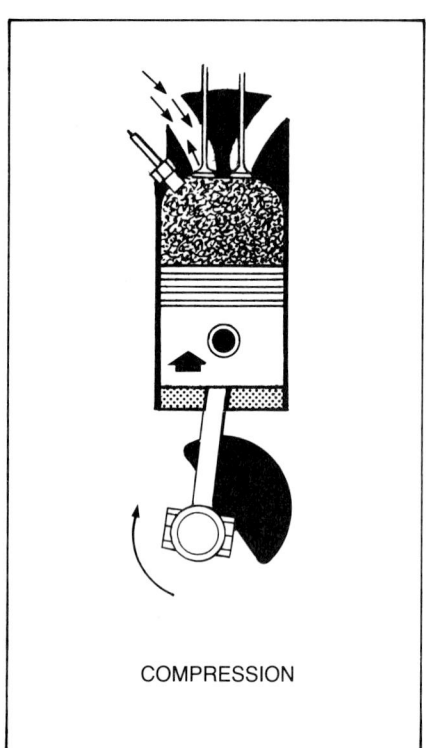

Closing intake valve causes incoming mixture to bounce back, creating a restriction to incoming flow.

Higher engine speeds caused a new set of problems in early engine designs because the effects of inertia became more significant. Because it wasn't mechanically driven, one of the first devices to reach its limitations was the automatic inlet valve. At higher speeds, air pressure just couldn't operate an automatic intake valve with the necessary timing precision for mixture control.

The inertia of the mixture column in the intake manifold became a major problem as engine speeds increased. An engine's intake manifold is basically a refined tube, with an intake valve at one end that pops up and down—they were called *poppet valves*—and a carburetor at the other.

As soon as the intake valve pops open, mixture should begin to flow into the cylinder. When the valve pops closed, the mixture flow should stop. But neither the valve nor mixture can stop moving instantaneously. We can change the opening and closing time and speed of a mechanically operated valve by changing camshaft profile and, if required, the valve spring. But we can't do too much to make the mixture flow change its response time. It could be possible to turn an engine so fast that the valves open and close before the mixture column has a chance to move. The engine would be barely breathing.

Considered from one angle, an engine is an air pump—mixture pump—for our purposes. The more mixture that can pass through it in any given time, the more horsepower it can produce. An engine's pumping efficiency is called *volumetric efficiency (VE)*.

VE is a measure of how well an engine breathes: The better the breathing capability, the higher the VE. It is defined as the ratio of the actual mass of mixture taken into the engine to the mass the engine displacement would theoretically consume if there were no losses. The ratio is expressed as a percentage. VE is low at idle and low rpm because the engine is throttled. It reaches a maximum at an engine speed close to where maximum torque at wide-open throttle (WOT) occurs; then it falls off as engine speed is increased to peak rpm.

For example, a 100 cubic inch displacement (CID) engine that takes in 80 cubic inches of mixture has a VE of 80%. Ordinary low-performance engines have a VE of about 75% at maximum rpm and high-performance engines have a VE of about 80% at the same condition. Full-race engines have a VE of about 90% or higher at their peak rpm.

What keeps an engine's VE below 100%? Well, the mixture must travel rapidly (when the engine is at higher rpm) through bends and constrictions in the carburetor and intake manifold. And it is exposed to heat from the engine and sometimes the exhaust manifold. The rapid movement and heat limit the amount of mixture that can go into the cylinder. The cylinder is only partially filled at higher rpm.

Webers can improve VE by reducing the restriction in intake mixture flow caused by the carburetor. Other restrictions include: air cleaner, intake manifold design, intake valve guide, intake valve and valve timing. On the exhaust side of the engine, poor VE can result from bad exhaust system design. But for now, we're interested in the loss of VE caused by the inertia of the mixture column in the intake manifold. To explore that, let's return to the Otto cycle and examine mixture flow in more detail.

Mixture Dynamics—Let's begin just before the intake stroke, when the intake valve is closed: No mixture is moving, because there's no place to go.

Now, as the intake valve begins to open and the piston starts down on the intake stroke, the mixture in the manifold is pushed into the cylinder because of atmospheric pressure. The mixture's momentum helps fill the cylinder, and its velocity increases as the piston reaches its maximum speed, about halfway down the cylinder. As the piston nears BDC it begins to slow, slowing also the flow of mixture into the cylinder.

The intake valve closes shortly *after* the time the piston starts back up on the compression stroke. That's news to the incoming mixture in the intake manifold; it's moving toward the cylinder because of its momentum. As a result, some of the mixture in the intake manifold bounces off the closed intake valve and starts back as a wave that works against the incoming mixture flow. As a result, the incoming mixture tends to lose momentum.

What happens as the exhaust valve begins to open? Burned mixture is pushed out into the exhaust manifold by the pressure remaining in the cylinder and the motion of the ascending piston. The burned mixture has inertia, too, so it keeps moving even after the speed of the piston slows as it nears TDC. This causes a *scavenging effect* that helps get the burned mixture out of the engine. This effect helps draw a new mixture charge into the combustion chamber when the intake valve opens near TDC of the exhaust stroke.

Let's go over the process again, but this time examine what happens as en-

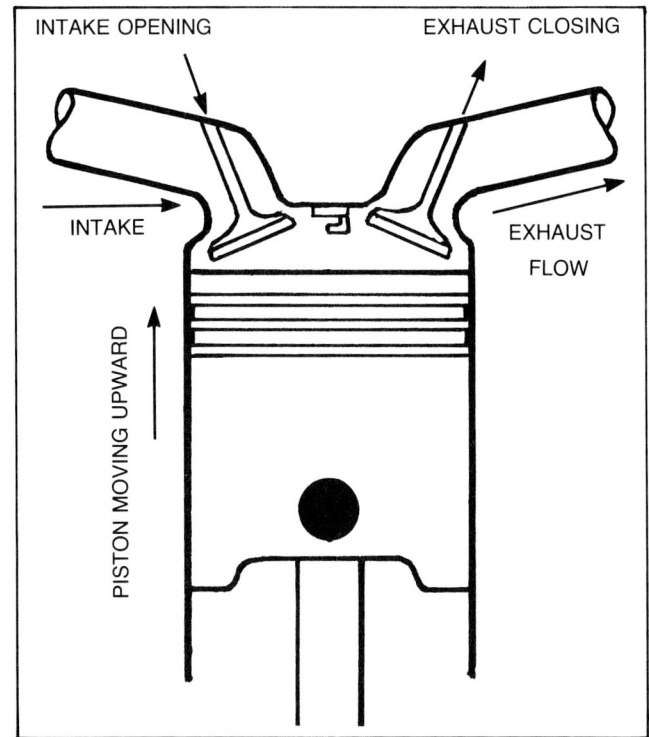

By opening intake early and closing exhaust late, both valves are open for a period called *overlap*.

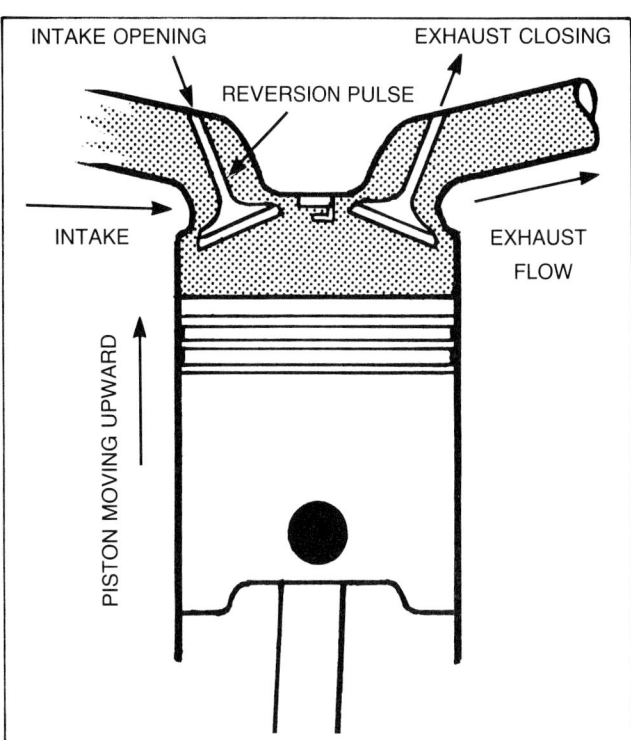

Reversion is caused when exhaust gases team with natural pulses in inlet manifold to force gas *out* of carburetor.

gine rpm increases. Because the valves and piston are mechanically operated, they can change timing and direction much faster than the mixture in the engine. Unless we fix this condition, VE will suffer and the power output of the engine will not be up to its potential.

The fix is to open the intake valve earlier and close the exhaust valve later to take advantage of the time it takes the mixture to overcome inertia and change momentum. Thus, the intake valve can begin to open several degrees before the piston actually reaches TDC on the exhaust stroke. Because we're dealing with high rpm, the early opening gives the intake valve a chance to be well open by the time the mixture can start flowing into the engine.

In a similar way, we can open the exhaust valve before the piston reaches BDC of its power stroke. The pressure inside the cylinder causes the exhaust to start flowing in a kind of shock wave that expands past the opening exhaust valve and travels out the exhaust port and manifold. This wave helps carry the exhaust in its lower-pressure wake. We can delay exhaust valve closing even after the piston has reached the top of the stroke because the momentum of the exhaust will continue to draw it out of the cylinder.

Because both intake and exhaust valves are open at the same time, the momentum of the exhaust helps start the flow of the intake charge into the cylinder. The period that both valves are partially open is called *overlap*.

Overlap is restricted in a slow-speed engine because the inertia of the mixture flow can keep up with the mechanical movement. In a high-speed engine, overlap assists the mixture in moving into and out of the cylinder in the desired time.

It's possible for exhaust to flow back past the intake valve with excessive overlap at part-throttle opening and low rpm. How? Assume the exhaust valve is still open and the exhaust is flowing out when the intake valve opens. At low rpm and part throttle, intake manifold pressure is less than atmospheric, and cylinder pressure is also higher than in the intake manifold. So the opening of the intake valve establishes a path for the exhaust gases to travel from a higher pressure area (in the cylinder) to a lower pressure area (the intake manifold).

This is called *reversion*, and can occasionally be seen as a plume of mixture blowing out of the carburetor when the throttle is suddenly opened wide. This burned exhaust gas can't be reburned, so it takes up space in the intake manifold and *dilutes* the incoming mixture. Reversion can indicate poor valve timing or a poorly designed intake and exhaust system. Or it can be a natural byproduct of an engine set up to operate only at WOT and high engine speeds.

Now, let's add some cylinders to a high-speed engine. We've already noted that there are pulses in both the intake and exhaust passages caused by mixture inertia, and these pulses act like waves. By

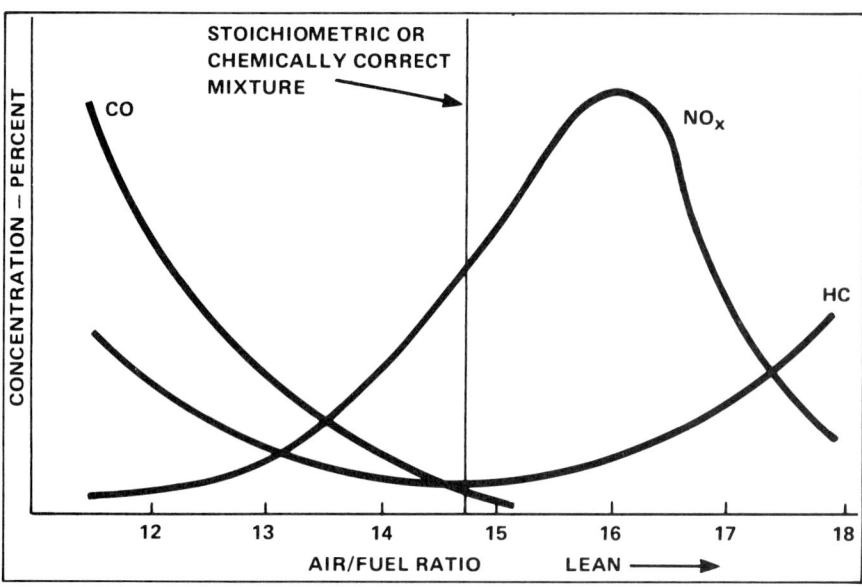

Approximate relationship between CO, HC and NOx as air/fuel ratio changes.

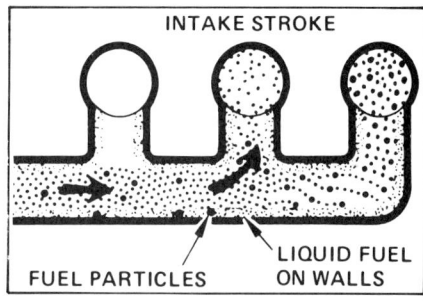

If mixture is partially vaporized, liquid particles clinging to manifold walls—or avoiding sharp turns into cylinder—may cause some cylinders to run lean and some rich. Here, left cylinder tends to run lean and end cylinder receives rich mixture.

changing the length or diameter of the intake or exhaust runners, the natural pulses in either manifold can be used to assist mixture flow through the engine. Changing, or "tuning," the intake or exhaust manifold for maximum power is a subject for another book.

One final note about intake system *dynamics.* The start of mixture flow past an intake valve is a *signal* to the carburetor to start feeding more air/fuel mixture. This signal becomes stronger as the carburetor is moved closer to the intake valve and weaker as it is moved farther away. For smooth operation, a weak signal is desirable. But for quickest throttle, a strong signal is essential.

Mixture Strength—Let's consider a single revolution of an engine at low rpm. The objective is to fill the cylinder with just enough gasoline mixed with air so that the gasoline will completely burn. This ideal mixture is called *stoichiometric.* It represents a combination of 14.7 parts air to one part gasoline by weight and is expressed as a ratio, 14.7:1. An engine with perfect VE burning a stoichiometric mixture should develop as much horsepower as is theoretically possible. As already discussed, this ideal isn't usually achieved in a production engine. As a result, efficient carburetion and low exhaust emissions involves juggling mixture richness to maintain the 14.7:1 ratio.

Gasoline burns best when it is broken up into very small droplets, each one of which is surrounded by a mist of gasoline vapor. Gasoline in that condition is *atomized,* and one of the carburetor's primary tasks is to atomize the fuel into a state that's something like the mist that comes out of a perfume bottle.

A carburetor does a fine job of atomizing fuel, but fuel doesn't stay perfectly atomized on its way to the cylinder. For the most part, it stays as a bunch of droplets of liquid carried along by air. The liquid gasoline always has greater mass and, thus, greater inertia, than the air it travels with. So the gasoline behaves differently. It turns corners slower than the air, drags along the inside surfaces of the manifold runners, and settles (condenses) into pools on the bottom of the manifold plenum and runners.

Liquid gasoline won't burn efficiently in an engine. Any pools of gasoline in the intake manifold are wasted, even if they dribble into the cylinder. Even worse, the gasoline that falls out of the intake charge into liquid changes the air/fuel ratio. The mixture becomes excessively lean and the engine loses power.

Inertia and condensation both cause the mixture to lean. As a result, most of the extra devices on a carburetor are designed to provide a richer mixture. Extra gasoline has to be atomized into the intake manifold to make sure the mixture finally reaching the cylinder is near stoichiometric. If emissions are not a concern, enough gasoline can be added so the engine is just short of fouling out the spark plugs. Racing engines run wonderfully rich. Modern street engines, however, need to strive for a mixture nearing stoichiometic for the best compromise of power, economy and emissions control.

A very simple carburetor can supply a stoichiometric mixture for an engine that runs at constant speed. An example is a lawnmower carburetor, which is anvil-simple and has a single jet. Weber carburetors are more complex. Let's examine their design and construction next.

WEBER DESIGN ELEMENTS

Venturi—Again, the primary purpose of the carburetor is create an atomized spray of gasoline so the engine is supplied with a nearly stoichiometric mixture. When liquid is exposed to air pressure lower than atmospheric (14.7 psi), and a low-pressure area is created by necking down the passage through which incoming air flows, it vaporizes more easily. And a low-pressure area is created in a carburetor with a *venturi*.

Exactly how this physical property works was the subject of considerable research toward the last part of the 18th century. The big name in this business is Giovanni Battista Venturi (1746-1822), an Italian professor from Modena who tried to figure out how fluids flow when a tube has different diameters.

He discovered that, if you reduce the diameter of a section of tube, fluid flows faster through the reduced section. When fluid velocity increases in this constriction, pressure decreases there. The constriction is called a venturi.

Carburetors are basically *pressure-differential* devices: The accompanying illustration shows what happens to pressures along the bore of the carburetor.

The key to understanding how a carburetor works is to understand that there is lower air pressure at the reduced section (venturi) of a tube than at its main diameter. All that is needed is to tap into this low-pressure point with a tube connected to a pool of gasoline. Atmospheric pressure will force gasoline through the tube and into the air stream, where it is atomized in the higher-velocity, low-pressure area of the restriction.

There are other design requirements for a carburetor. One is the control of fuel flowing to the venturi, because there isn't a valve that's turned on and off to start and stop its flow. If the tip of the nozzle in the venturi were positioned below the surface of the gasoline, then gasoline would dribble continuously into the manifold. On the other hand, if the nozzle were placed very high above the gasoline's surface, the weight of gasoline in the tube would counteract the air pressure pushing it into the venturi and no gasoline would flow.

The vertical relationship between the surface of the gasoline pool and the nozzle in the venturi has a significant effect on the amount of gasoline delivered.

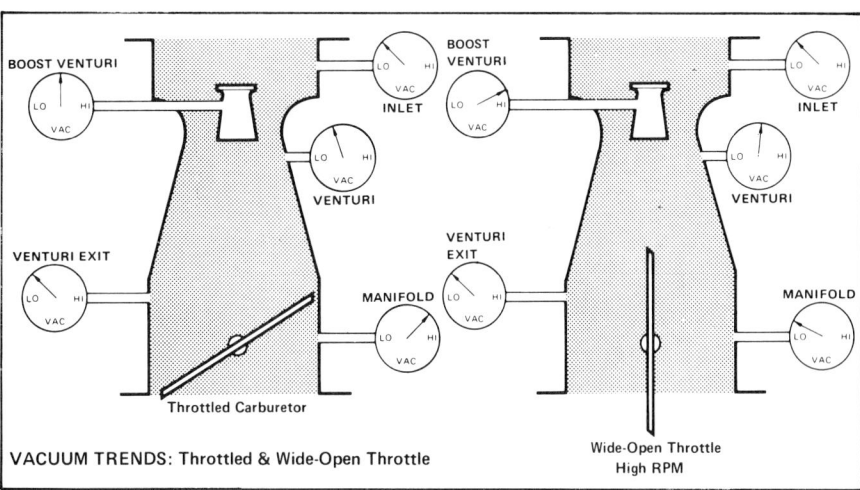

Typical vacuum drops inside carburetor: Slight vacuum at INLET represents pressure drop across air cleaner. Gage at VENTURI throat show higher vacuum because air is still at relatively high velocity. Vacuum returns almost to INLET value just before throttle plate at VENTURI EXIT. Throttled carburetor shows very high MANIFOLD vacuum because a large pressure drop occurs across the partially-opened throttle. WOT low MANIFOLD vacuum reading indicates a heavy load/dense charge condition. In both cases, highest carburetor vacuum is at BOOST VENTURI supplying signal to main system.

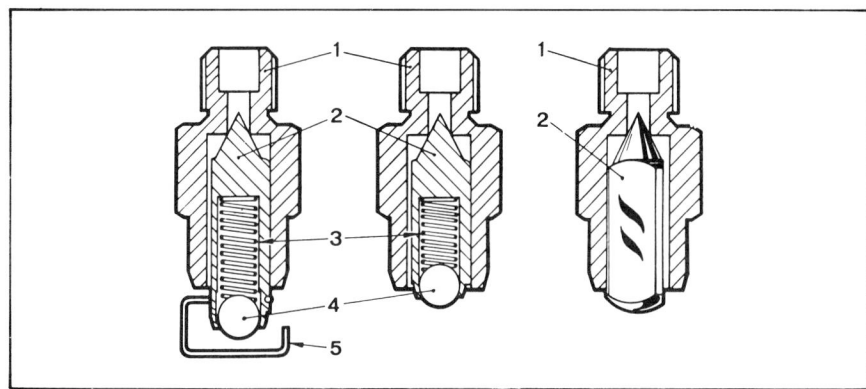

Needle valves: (1) Needle-valve seat, (2) needle valve, (3) damper spring, (4) damper ball, (5) needle-valve hook. Drawing courtesy Weber.

This, in turn, influences the air/fuel ratio of the combustible mixture. In practice, the nozzle is placed slightly above the pool's surface to minimize the weight of fuel in the passage to the nozzle.

Float & Needle Valve—The *fuel bowl* or *float chamber* is the reservoir supplying fuel in the carburetor. Fuel level in the chamber is controlled by the *float* and *fuel inlet valve* or *needle valve*. Maintaining a constant fuel level in the float bowl is a critical carburetor operation. This level determines the fuel level in carburetor's other passages. A fuel level too high will create too rich an air/fuel mixture; too low will produce one that's too lean.

How do the float and needle valve

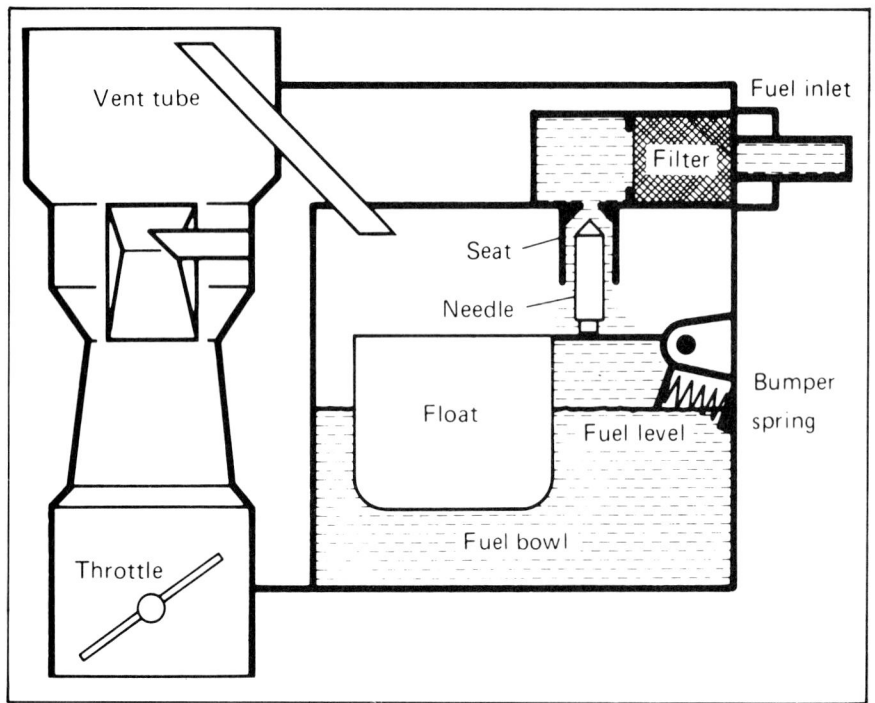

Typical inlet system: Float hasn't reached desired level, so needle valve admits fuel to bowl. Fuel flow shuts off when float rises to close needle against seat.

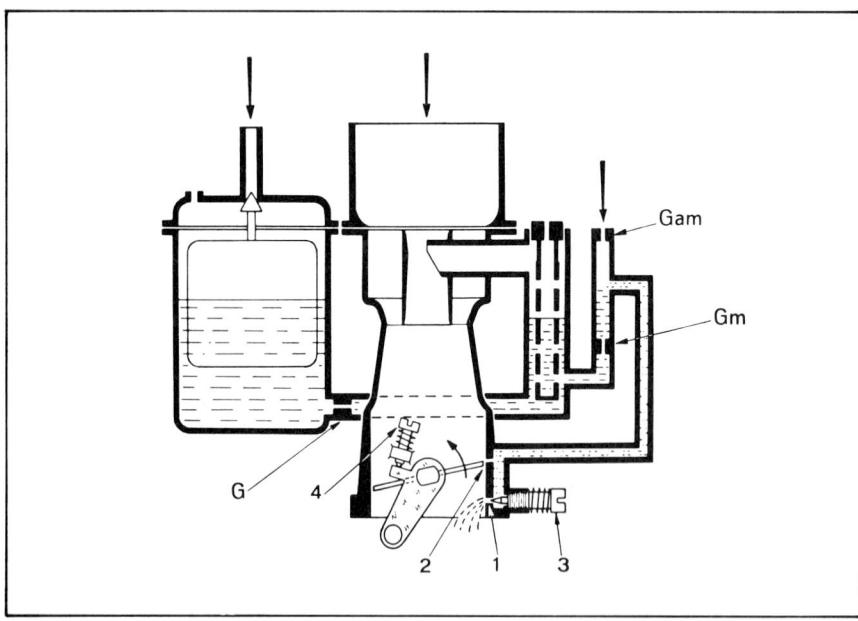

Idle circuit supplies fuel below throttle plates. Mixture richness is adjusted by needle valve as well as idle jet: (Gam) idle-speed air jet, (Gm) idle-speed fuel jet, (G) main jet, (1) idle-mixture hole, (2) transition (progression) hole, (3) idle-mixture screw, (4) idle-speed screw. Drawing courtesy Weber.

work to keep a constant fuel level? When gasoline is extracted from the fuel bowl, the float lowers, or drops. Fuel-pump pressure forces the needle valve open and more fuel enters the bowl. The fuel level rises, the float does, too, and it closes the needle valve; fuel flow into the bowl stops. The needle valve's opening and closing is rapid and the fuel level remains almost constant.

Here are a few tips about the float system. Too high a fuel pressure to the carburetor can push the needle valve off its seat and cause an excessively rich air/fuel mixture and possibly flooding. What results? Miserable driveability at best, an engine fire at worst. In general, Webers should receive a maximum fuel pressure of 3.5-psi. Many standard mechanical and electric pumps can easily exceed this. Buy a *fuel pressure regulator* and install it in-line just before the carburetor(s).

Many, many carburetor problems are caused by dirt somehow reaching the internals. Install a quality, low-restriction, in-line fuel filter ahead of the pressure regulator. And keep it changed!

CARBURETOR CIRCUITS

When engine rpm increases, maintaining a good mixture becomes increasingly difficult. Because gasoline has more mass than air, it responds differently to pressure changes resulting from different engine speeds. Gasoline has 557 times the mass of air, so the difference in the way gasoline and air flow during pressure and direction changes is significant.

A modern carburetor uses two different solutions to address this problem. The first is to segment engine speeds handled by the carburetor, and divide them into different circuits that can be tuned to provide the desired air/fuel mixture. The other solution is to reduce the mass of the gasoline so it flows with almost the same characteristics as air. This is accomplished by bubbling air into the fuel as it flows through circuits inside the carburetor.

Idle Circuit—At idle, very little air flows to the engine past the almost-

CONSTANT-VELOCITY CARBURETORS

While this is a diversion from Weber carburetor design, it's useful to see that there are other ways of solving the problems of maintaining the correct air/fuel mixture to an engine.

G.B. Venturi discovered: (1) a restriction in a tube causes fluid flowing through it to increase in velocity at the restriction (the venturi), and (2) fluid pressure at the restriction is less than pressure in the rest of the tube. Venturi's discovery is the basis on which carburetors operate.

As engine speeds increased, it was found that liquid gasoline flows more readily than air. The net result is that, at higher engine speeds, an uncorrected circuit flows increasingly more gasoline than air.

Now, if you change the restriction size to match air flowing past it, you'd be able to have a *constant-velocity* device that would need only a fixed air/fuel mixing jet to supply a stoichiometric mixture. In other words, instead of switching to other circuits with different jet sizes—as in Holley, Carter, Solex, Weber and similar carburetors—you change the size of the restriction—as in SU, Amal and Bing carburetors. One advantage with this type of carburetor is that it uses a single jet.

Skinner Union (SU) carburetors were standard on many British engines produced before the '70s. The 1970—72 Datsun 240Z used Hitachi built versions of SUs. They operate using manifold pressure to vary the venturi in the main throat of the carburetor. As venturi pressure drops (as the engine accelerates), the venturi size is enlarged, and a constant mixture flow velocity is maintained at the restriction. The main fuel jet is placed at the venturi, just as in a conventional fixed-jet carburetor.

The SU main jet always encounters the same velocity of air; as a result, only a minor correction to fuel flow is required to match the required air/fuel mixture. The correction is done by a tapered needle, attached to the sliding piston, which works in and out of the main jet.

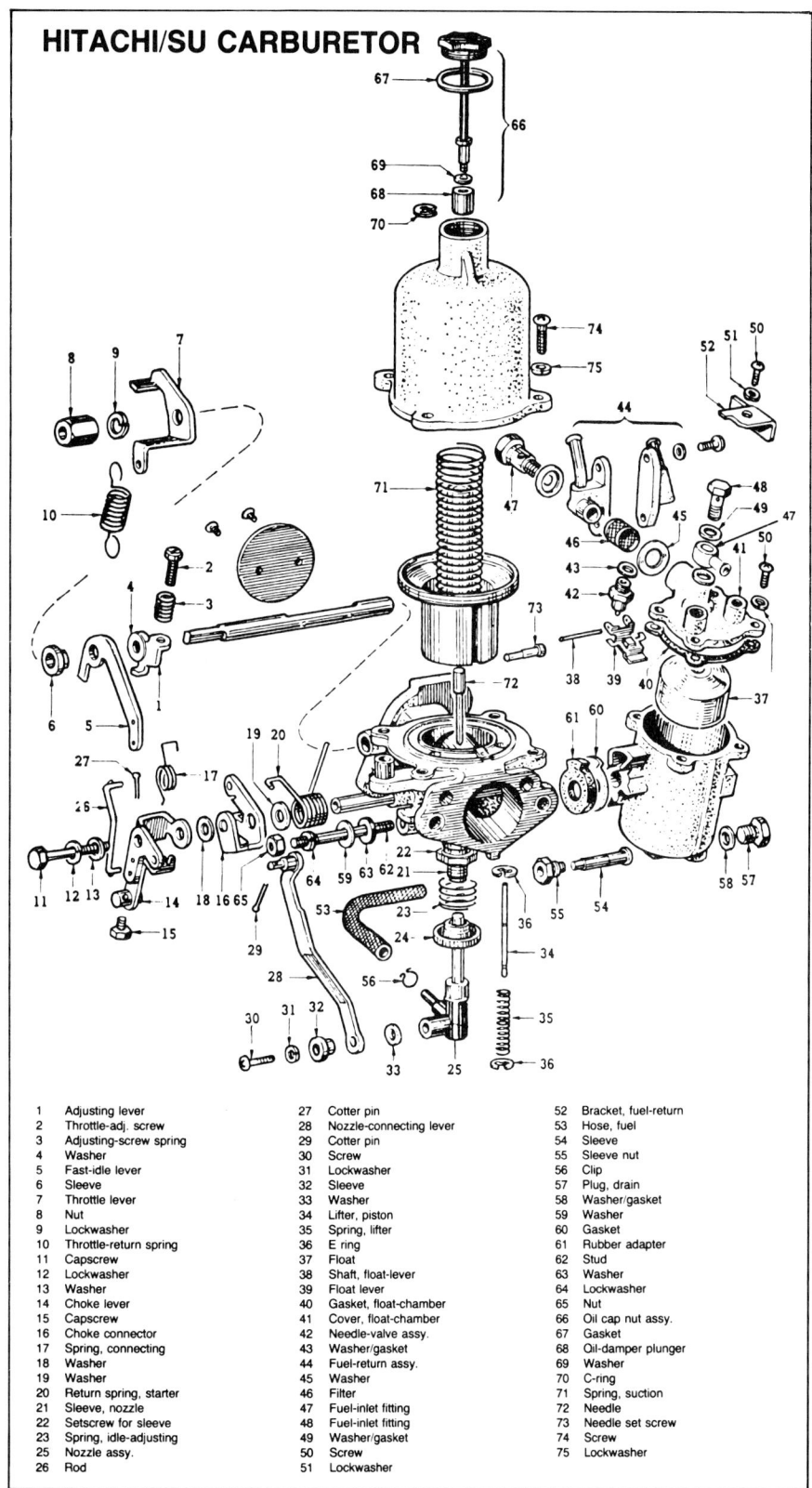

HITACHI/SU CARBURETOR

#	Part	#	Part	#	Part
1	Adjusting lever	27	Cotter pin	52	Bracket, fuel-return
2	Throttle-adj. screw	28	Nozzle-connecting lever	53	Hose, fuel
3	Adjusting-screw spring	29	Cotter pin	54	Sleeve
4	Washer	30	Screw	55	Sleeve nut
5	Fast-idle lever	31	Lockwasher	56	Clip
6	Sleeve	32	Sleeve	57	Plug, drain
7	Throttle lever	33	Washer	58	Washer/gasket
8	Nut	34	Lifter, piston	59	Washer
9	Lockwasher	35	Spring, lifter	60	Gasket
10	Throttle-return spring	36	E ring	61	Rubber adapter
11	Capscrew	37	Float	62	Stud
12	Lockwasher	38	Shaft, float-lever	63	Washer
13	Washer	39	Float lever	64	Lockwasher
14	Choke lever	40	Gasket, float-chamber	65	Nut
15	Capscrew	41	Cover, float-chamber	66	Oil cap nut assy.
16	Choke connector	42	Needle-valve assy.	67	Gasket
17	Spring, connecting	43	Washer/gasket	68	Oil-damper plunger
18	Washer	44	Fuel-return assy.	69	Washer
19	Washer	45	Washer	70	C-ring
20	Return spring, starter	46	Filter	71	Spring, suction
21	Sleeve, nozzle	47	Fuel-inlet fitting	72	Needle
22	Setscrew for sleeve	48	Fuel-inlet fitting	73	Needle set screw
23	Spring, idle-adjusting	49	Washer/gasket	74	Screw
25	Nozzle assy.	50	Screw	75	Lockwasher
26	Rod	51	Lockwasher		

SU carburetors were standard equipment on 1970—'72 240Zs. Drawing courtesy Nissan.

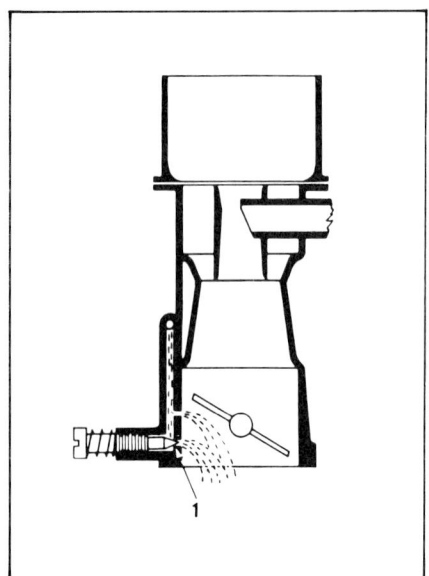

Progression holes supply fuel in just off-idle transition. In actual practice, there may be three to five holes that are opened as throttle plate rises. Drawing courtesy Weber.

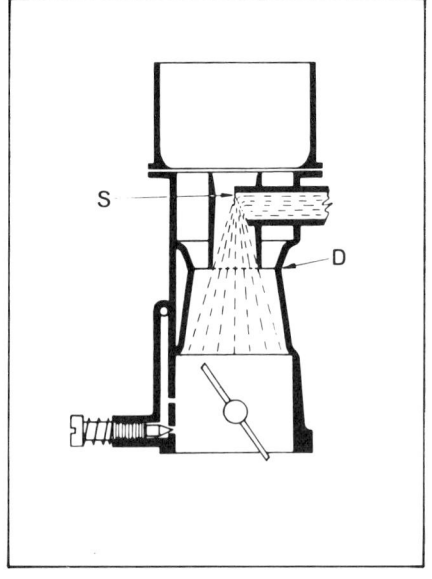

Main circuit sprays fuel into air stream at narrowest section of venturi where air velocity is highest, pressure lowest. (S) is spray nozzle and (D) is venturi. Drawing courtesy Weber.

closed throttle plate, and the engine requires very little fuel. In fact, airflow is so slow that there isn't enough velocity through the main venturi to "pull" gasoline into the circuit.

The throttle plate itself, however, can also act as a venturi when it is almost closed, and there is a workable pressure drop immediately downstream of the throttle plate at idle. This pressure drop is used to draw fuel through the *idle circuit*.

The amount of fuel drawn through is regulated by a needle, usually called the *idle mixture adjustment screw*. Turning it in (clockwise) restricts the passage and leans the mixture. Turning it out (counterclockwise) opens the circuit and richens the idle mixture.

Fuel flows through the idle jet, then passes through another jet where it is mixed, or *emulsified*, with a small amount of air. As the air bubbles along with the fuel, it reduces the total mass of the fluid so it more closely matches the flow of air and maintains the correct air/fuel mixture.

As long as the throttle plate is closed (the engine is idling), the idle circuit is the only source of fuel to the engine. This assumes the choke isn't activated. I'll get to that in a moment.

Progression Circuit—As the throttle plate begins to open, the low-pressure area travels slightly upwind in the carburetor, following the edge of the throttle plate, and begins to draw fuel through a series of holes in the side of the throttle bore.

In some Weber carburetors, these holes are cut *after* the throttle plate has been assembled to its shaft. These holes are custom-matched to the position of the individual throttle plate.

Why such attention to a detail? Because the positioning of the *progression holes* is critical in smoothing the transition between closed and open throttle. If the placement of the holes is not correct, the engine stumbles and surges coming off idle, and driveability suffers.

When the throttle plate opens past the progression holes, engine rpm increases and air speed past the throttle plate increases. This increase further lowers the pressure just downstream of the edge of the throttle and draws more fuel into the airflow.

The fuel continues to be drawn through the idle jet, and more fuel is delivered by the progression circuit as the throttle plate passes the several holes drilled in the side of the throttle bore.

As the throttle opens farther, airflow begins to be high enough so that the main fuel delivery circuit starts to register a pressure drop at the main venturi and fuel begins to flow from the *main circuit* (see below).

It's just as well, for the throttle plate can open only so far before the gap between the edge of the plate and the throttle bore is so large that there is no longer a sufficient pressure drop at the gap.

In fact, as the throttle opens farther, the flow of fuel through the idle and progression circuits tapers off, because the pressure drop is lost. It's a challenge to design a carburetor that can switch smoothly between the lower-speed idle and the higher-speed main circuit. It must continue to provide the correct air/fuel mixture during the transition.

Main Circuit—By about 1500 rpm, airflow velocity through the venturi is high enough to produce a pressure drop that causes gasoline to begin flowing through the *main circuit*. Like the idle/progression circuit, the main circuit includes an air-bubbling correction to manage fluid flow over a range of speeds.

Webers, like many other carburetors, use two venturis. One, the *main venturi*, nearly approaches the throttle bore in diameter, and another, the *boost* or *auxiliary* venturi, is centered in the main venturi. This auxiliary venturi is positioned in the center of the bore and is where the fuel is actually atomized.

The main venturi supplies the primary pressure drop in the carburetor. The tail of the auxiliary venturi ends at the narrowest section of the main venturi where the pressure drop is most pronounced.

The auxiliary venturi offers only a slight necking down compared with the profile of the main venturi, but it does make the main circuit more sensitive to the pressure drop supplied by the main venturi. And, it positions the fuel outlet squarely in the center of the bore.

Auxiliary Circuits—As precise as car-

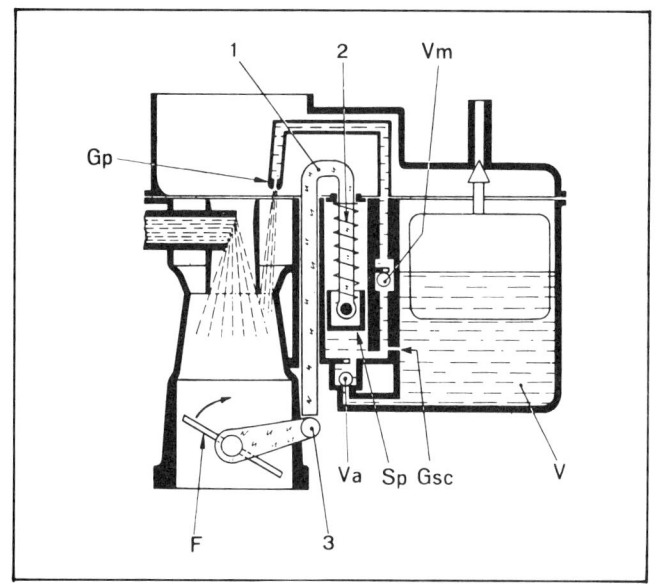

Piston-type accelerator pump. Pump plunger (Sp) is controlled by arm (3) connected to throttle valve. Drawing courtesy Weber.

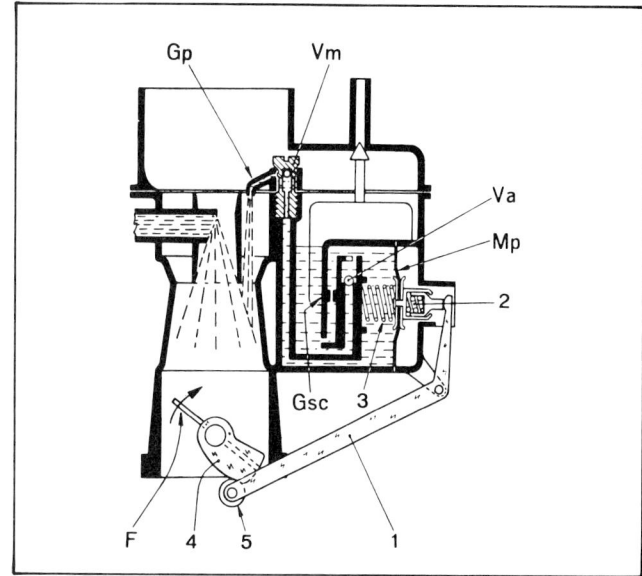

Diaphragm-type accelerator pump. Drawing courtesy Weber.

buretor circuits are, they still fail to supply the needed stoichiometric mixture to the engine at all times. Two instances are handled by special circuits: sudden acceleration and cold starting.

Accelerator Pump: When you suddenly "floor it," the engine experiences a somewhat unusual condition. The throttle plate is open all the way, but engine speed is relatively slow. Because the engine is pumping slowly (low rpm) through a big hole (WOT), pressure in the intake manifold approaches atmospheric. This condition drops the airflow at the venturi so low that the air/fuel mixture goes instantly lean. This excessively lean mixture causes a hesitation, or *flat spot*.

Now, the engine will gradually correct this condition. As soon as rpm increases, airflow increases and the venturi will start working again. But in that instant when the throttle plate is fully open, the engine needs an extra dose of fuel to keep the air/fuel mixture combustible.

The extra dose is provided by the *accelerator pump*, which squirts a shot of raw gasoline into the intake manifold when the throttle is opened. On a DCOE carburetor, the accelerator pump is literally a pump: a piston in a cylinder. The piston is operated by a cam and linkage off the throttle shaft. On other Webers, the pump is a diaphragm (like in a mechanical fuel pump), which acts in essentially the same way as a piston pump in delivering fuel.

A series of one-way check balls keeps fuel flowing in the right direction. On a DCOE, these checks are little balls that lift off a seat when fuel is pumped in the desired direction, and reseat when the pump starts to force fuel the wrong way. In other carburetors, the balls are replaced by flaps of rubberized material, or spring-loaded valves.

Starter Circuit (Choke): A cold engine is another instance when the carburetor can't deliver the correct mixture. In this case, it's not so much because the engine is running too slowly. It's more the result of the engine's cold intake manifold, which causes fuel to *puddle*—condense out into pools. It's the same phenomenon that causes windows to frost in winter, or ice to form on the coils of a refrigerator. This is not, however, *carburetor icing,* which is caused when water vapor freezes and blocks the main fuel feed in the carburetor.

There are really two kinds of starting devices. The most common one is the

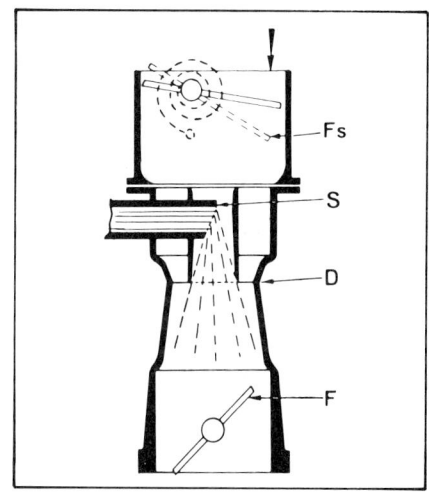

Typical *choke plate* or *valve* (Fs) literally "chokes" engine (British term is "strangler"), producing low-pressure/high-velocity flow of air to pull fuel through main circuit. Drawing courtesy Weber.

butterfly valve, or *choke,* which sits at the very top of the throttle bore. When it is almost closed, the *choke plate* creates a major restriction "choking" the incoming air. When the engine is cranked, the pressure drops—there's a partial vacuum—under the plate and more fuel in pulled from the main circuit. This is enough for a starting mixture.

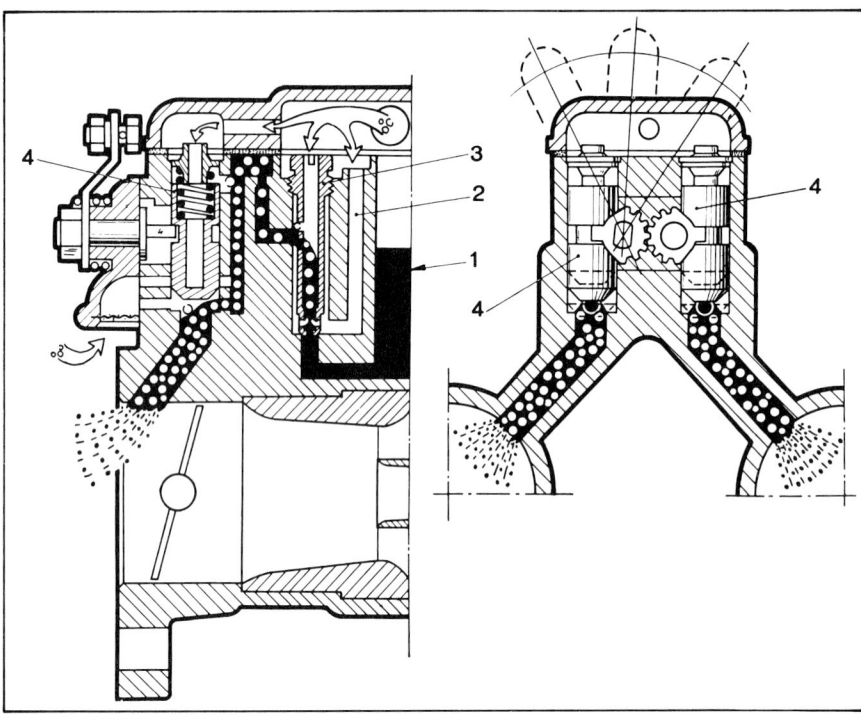

DCOE choke-system schematic: (1) Float chamber with fuel, (2) starting fuel-reserve well, (3) choke jet with emulsion tube and air jet, and (4) plunger. Drawing courtesy Weber.

POWER VALVE VALUES

Note that values of valves, as with most Weber parts, have no logical relation to part numbers. Values under *Opens* are in millimeters of Hg and arranged as minimum/maximum.

P/N	Opens
57804.025	127/165
57804.052	190/228
57804.096	152/177
57804.097	165/203
57804.123	89/127
57804.220	63/88

Weber is a bit more sophisticated in the starting circuit of the DCOE and some other carburetors. There is a separate fuel circuit, complete with air correction, which is intentionally tuned very rich. This circuit is turned on and off by sliding valves that control both air and fuel flow. The excessively rich mixture is delivered through a rather large hole in the throttle bore downstream of the throttle plate. In these carburetors, there is no choke plate. Instead, Weber has what is literally a carburetor within a carburetor to supply the extra-rich mixture required to start a cold engine.

Power Valve: The power valve is a modern necessity. To keep emissions low, most engines produced since 1971 for use in the U.S. are tuned to run lean. In the discussion about the accelerator pump, I noted there are instances when the engine needs more fuel. Accelerator pumps handle the problem nicely when going from very low to higher engine speeds, because the rotation of the throttle shaft is great enough to make a cam or some kind of mechanical linkage work the pump.

At high engine speeds, however, the throttle is almost fully open, and there isn't enough throttle shaft movement (rotation) left to operate an accelerator pump. Yet, the engine needs a little richer mixture in the 3000—5000 rpm range, too. This need is detected by the vacuum-activated *power valve,* which is a spring-loaded diaphragm set to open under low-vacuum, high pressure conditions in the intake manifold. When manifold pressure rises (vacuum drops), a spring opens the valve and supplies addtional fuel to the main system during high-speed operation.

EMISSION CONTROL DEVICES

For many enthusiasts, emission control devices are an anathema. Their hatred could have originated because auto manufacturers chose a band-aid approach to emissions control in the post-1971 era when the federal law was new.

Mixtures were leaned to reduce hydrocarbon (HC) and carbon monoxide (CO)

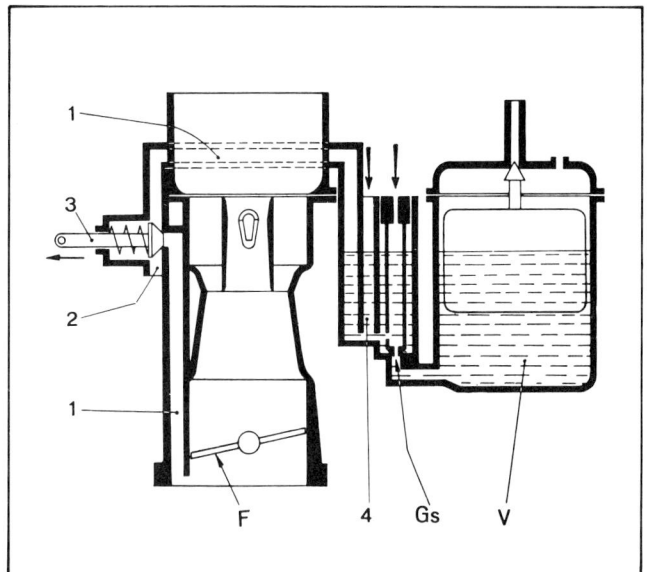

Weber starter circuit opens separate—very rich—fuel-flow path through carburetor. Advantages: main venturi isn't blocked, sudden full-throttle acceleration is unrestricted with cold engine. (1) Starting mixture duct, (2) starting air jet, (3) starting valve, (4) starting reserve well, (F) throttle, (Gs) starting jet, and (V) float chamber. Drawing courtesy Weber.

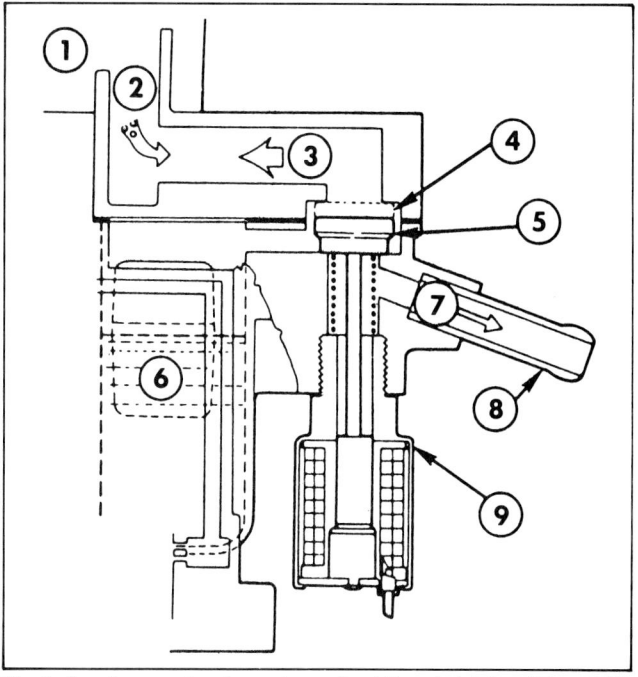

Float-chamber vent-valve schematic: (1) carb intake, (2) and (3) fuel vapor moves these directions with ignition on, (4) valve closes vent when ignition is off, (5) valve seat, (6) float, (7) fuel vapor vents to evaporative canister with ignition off, (8) charcoal-canister connection, and (9) vent-valve solenoid. Drawing courtesy Weber.

emissions, and that affected driveability. Exhaust gases were added—via exhaust gas recirculation or *EGR*—to dilute the mixture, cooling it and lowering oxides of nitrogen (NOx). But this influenced both performance and economy in the process. These early efforts were a mixed success. While they aided in stemming the spewing of emissions that are poisoning our planet, they didn't contribute to much driving enjoyment. Many people removed first-generation emission controls to get improvements in both driveability and economy.

Manufacturers have now produced clean-burning engines that drive well and return acceptable economy. As a result, removing emission controls on 1980s cars will have very little positive effect, and may even *rob* power, driveability and economy because the controls are so intimately integrated into the engines.

It's important to note that, after 1980, no original-equipment Weber carbureted cars were imported into the U.S. Cars that had Webers in Europe were fuel injected for this market. Alfa Romeo is an example.

It has been a real challenge for Weber distributors, who depend on the aftermarket conversion business for a lot of sales, to create Weber carburetor kits that improve performance and still retain compatibility with the emission control requirements. The problem is particularly critical in California, where bolting on a Weber to replace a stock Holley can be reason enough to fail the required bi-annual emissions inspection.

If California represented the potential of only a few carburetor conversions a year, the problem would not be critical. But the California market is one of the largest in the world, and certainly worth chasing. As a result, companies such as Redline have developed Weber conversion kits that are legal in California.

I don't delve into the chemistry that creates emission pollutants. But it is necessary to survey the devices that control them. Following are typical emission control devices used on Webers:

Electrically-heated choke with vacuum pull-off: An electrically heated choke operates independently of the operating temperature of the engine and functions over a shorter period of time and more precisely than the water-heated automatic choke. A *vacuum pull-off* opens the choke plate under low load/high manifold vacuum to keep from creating an excessively over-rich fuel mixture when the car is decelerating with a closed throttle.

Fuel cut-off solenoid: The solenoid blocks the idle fuel circuit when the ignition is switched off, and helps keep the engine from *running on* or *dieseling*.

Float-chamber vent valve: Gasoline vapors from the carburetor float bowl pose a health hazard. When the engine is running, the float bowl is vented to the air cleaner and fumes pass into it. When the engine stops, the vent passage to the air cleaner is sealed and vapor is directed to a *charcoal canister*. Fuel vapor is stored in the canister and then *purged*. It is drawn into the engine on restarting.

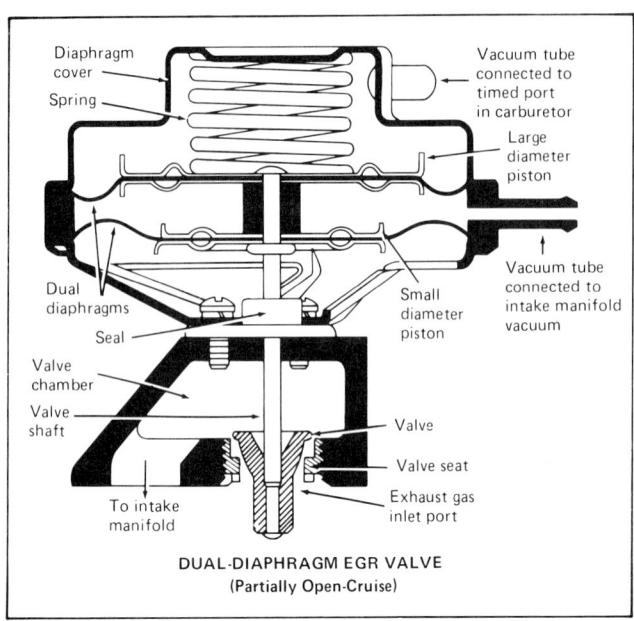

DUAL-DIAPHRAGM EGR VALVE
(Partially Open-Cruise)

Engine surging at light loads and part throttle is reduced by dual-diaphragm EGR valve.

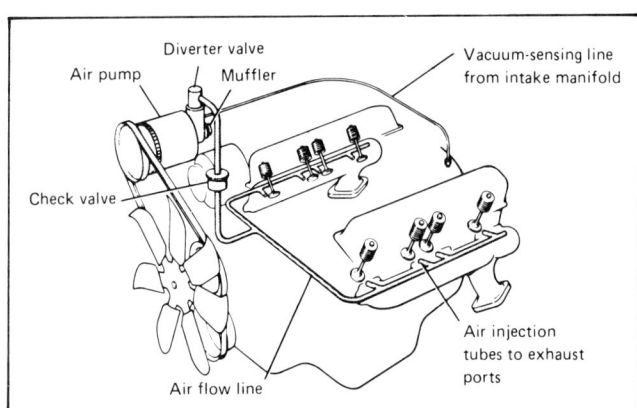

Air-injection systems are similar on all cars that use them. Chrysler version was first used in 1972 on California cars. Air pump adds controlled amount of air to ensure complete burning of exhaust gases in exhaust manifolds.

Fuel-return check valve: Many cars circulate gasoline in a loop that includes the carburetor and fuel tank. The idea is to fight *vapor-lock* by keeping the fuel cooled through constant circulation. The tank is a lot cooler than the top of an engine, where the fuel would possibly boil if it were not circulated. The check valve assures that fuel flows only one way in the system.

Sealed idle-mixture adjustment screw: This is a tamper-proof device to limit unwarranted adjustment of idle mixture richness.

RELATED SYSTEMS

Though not actually part of a carburetor, several systems are associated with emissions-reduction efforts.

Exhaust Gas Recirculation—EGR has a significant effect on what happens inside the cylinder. EGR reduces NOx by diluting the air/fuel mixture with a small portion of exhaust gases to lower combustion chamber temperature. This lower temperature helps prevent NOx.

Because the air/fuel mixture is usually a bit more lean than stoichiometric in typical production engines, combustion temperatures are typically high. Adding more fuel cools combustion temperature, but this also increases HC emissions. Consequently, another method was needed to cool the combustion process.

By recirculating about 6–10% of burned exhaust gases back into the air/fuel mixture, the combustible mixture is diluted with non-combustible exhaust gas. Therefore, there is a smaller, cooler "fire," and NOx is reduced.

EGR isn't needed or useful at idle because the amount of NOx formed is small. Maximum EGR usually occurs during cruising and accelerating between 30–70 mph, when NOx is likely to form. Each auto manufacturer has developed an EGR design. They accomplish the same task, but do clutter the engine department. Temperature sensors, flow valves and other hardware complicate troubleshooting efforts.

Air Injection—This method was one of the earliest used to meet emission standards. All the systems have the same basic design. A belt-driven *air pump* forces fresh air to *injector nozzles* in the exhaust manifold or cylinder head. This air mixes with the hot exhaust and helps burn leftover HC and CO in the exhaust to form H_2O and CO_2 (water and carbon dioxide). Air pumps are not major power robbers, but they are sometimes removed by the misinformed to increase power or uncomplicate the engine bay.

Catalytic Converter—This device is placed between the exhaust manifold and muffler. The catalytic converter uses the exhaust heat to work a chemical change in the exhaust gases and reduce emissions. *Two-way* converters are called *oxidation converters*: They remove HC and CO. *Three-way* converters reduce HC, CO and NOx and are called *reduction converters*. Three-way converters work best at reducing NOx when CO in the exhaust is between 0.8–1.5%.

Catalytic converters reduce the measured emission-controlled items, but add their own characteristic smell and some sulfuric acid to the atmosphere. Catalytic converters are not fragile, but certain operating precautions are necessary to keep them useful.

Engines with failed piston rings or ex-

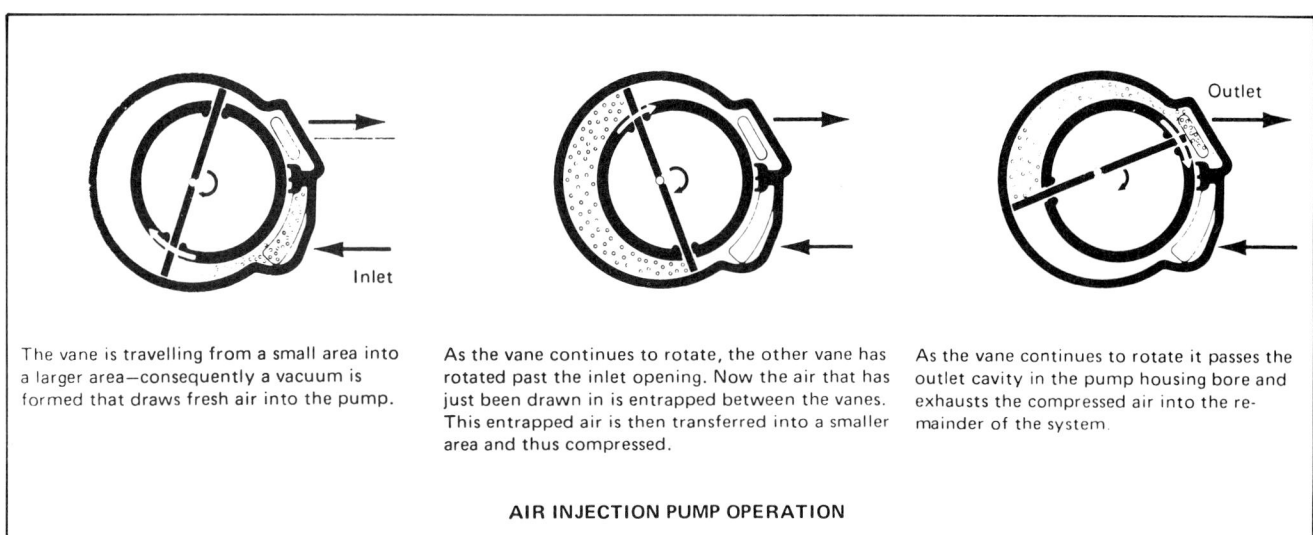

The vane is travelling from a small area into a larger area—consequently a vacuum is formed that draws fresh air into the pump.

As the vane continues to rotate, the other vane has rotated past the inlet opening. Now the air that has just been drawn in is entrapped between the vanes. This entrapped air is then transferred into a smaller area and thus compressed.

As the vane continues to rotate it passes the outlet cavity in the pump housing bore and exhausts the compressed air into the remainder of the system.

AIR INJECTION PUMP OPERATION

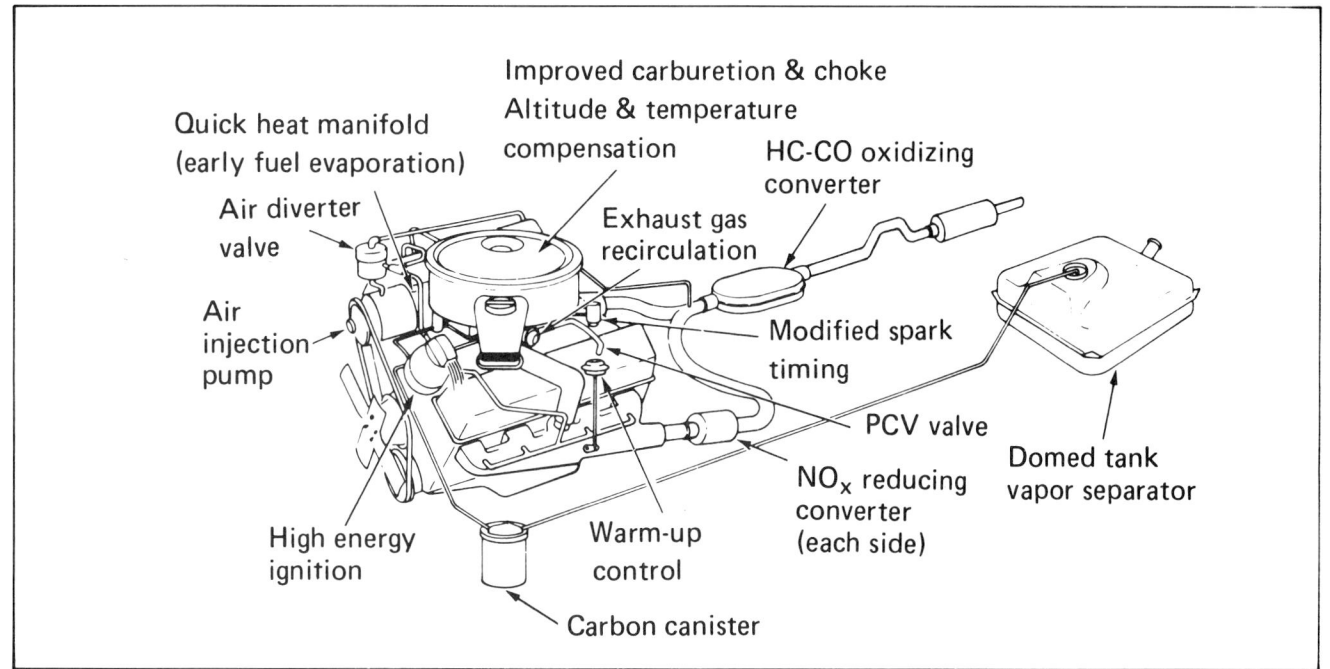

Emissions-control technology has progressed from simple PCV to complete package. Cost and complexity increased, too.

cessive HC output can damage the converter. So, correct engine tune and condition are important to converter life and performance.

Unleaded fuel *must* be used in cars with converters. Leaded fuel coats the catalyst and prevents exhaust gases from reacting with it.

Excessive engine temperatures can also destroy a converter's core. At temperatures higher than 1500F (816C), the catalyst breaks up or melts. Avoid idling a car with a converter for longer than 10 minutes.

In summary, the carburetor, like the internal combustion engine, started out as a simple device. It's relatively easy to construct a simple slow-speed internal combustion engine. But the demand for greater horsepower and driveability, coupled with clean air requirements, have made the engine and its carburetor more complex.

4
Troubleshooting

Weber "work-alike" carburetors offer Weber tuning versatility and adaptability to performance applications. These 40mm SK Racing carbs are mounted on an Oldsmobile Quad 4 for dyno testing. (Photo by Glenn Grissom, courtesy MSD Ignition.)

COUGH, COUGH

The odds of casually bolting on a high-performance Weber and driving away with a perfectly tuned engine are slim. After the installation, comes the troubleshooting.

Stories of the natural-born mechanic are legends—those who can understand engines as fluently as if they were conversing with their mother. Only a brief introduction, or none at all, is required for the "natural" mechanic to get right to the heart of what's troubling an engine.

Such people do exist. No doubt, many of them are making their living on the NASCAR or Formula 1 circuits, working on some of the most exciting internal combustion engines man can devise. No doubt, some of them spend much of their time in everyday jobs, dreaming of the NASCAR or Formula 1 engines that so badly need their talents.

This chapter is for the rest of us. I examine some of the tactics the accomplished mechanic uses. We may not strip away all the mechanic's mystery. But there are some skills that can help us understand how to become a better troubleshooter. And, that's what's needed, if you're going to select, install and tune Weber carburetors.

Experience—You can't become a champion weightlifter just by reading books about the subject. Neither can you learn about engines or carburetors without getting your fingernails dirty.

Getting experience means working on engines—all kinds of them. For the beginner, this usually means buying a used car and fixing it up. Mistakes made can be more valuable than the successes, because, very slowly, one develops an understanding of the limits: what is and is not possible in an internal combustion engine. Eventually, after all the snapped studs and torn seals are repaired, an old engine may begin to have new life.

Basic to all engine and carburetor troubleshooting is a knowledge of what something should be: what a healthy engine sounds like, what it feels like at idle or maximum throttle. This is similar to the weightlifter knowing what it feels like when clean-jerking a maximum weight. Once you know what it's like when it's right, you have a yardstick or baseline to measure against.

There are no substitutes for experience. To rephrase: There is no substitute for *intelligent* experience. You could live around engines all your life and, if you never paid attention, you'd know nothing about them.

Observation—So, experience is limited without observation, and here is where truly good mechanics originate. They use all their senses to try to understand engines. We usually think in terms of just sight and sound around an engine, but they have smells, too. This sense is more sensitive than sight or sound to warn when fuel is contaminated or overflowing, or oil is leaking onto a hot manifold. Furthermore, an engine has a rhythm that can be felt, and small errors in carburetion or ignition can produce intermittent misfires that can be sensed by laying your hand on a cooler part of the engine to feel its rhythm.

Clearly, sight is a most valuable sense around an engine. There are many stories of loose wires or backward installations that only need to be seen to be recognized. Yet, very little of what surrounds us is really seen. Skillful mechanics train themselves to see more of an engine than the ordinary person.

Logic—If we could see, hear, smell, taste or touch a problem right away, there'd be no need for diagnosis. One of the difficulties of troubleshooting is determining where to point our senses to find the source of a problem. That's where your brain is a powerful ally. Fortunately, diagnostic logic can be learned, and its principles are easy to explain.

Deduction—Sherlock Holmes used deduction to solve crimes. Deduction is the process of using general laws to reach specific conclusions. For example, Holmes sometimes thought like this: "As a general rule, any guilty man would do this. Jones did not do that, so he must not be guilty. That leaves Smith." In troubleshooting an engine, we use our experience, which has established some general laws, to draw new conclusions.

Some general laws about engines are: Correctly tuned engines idle smoothly, don't smoke and develop lots of power. The conclusion follows: This engine doesn't do any of these (it idles rough, smokes and develops little power); *therefore*, it must be out of tune. You can see how both experience and observation are critical to the process: Experience establishes the laws, and observation gives us the facts from which we deduce the conclusions.

An effective troubleshooter eliminates the good items until only the bad ones are left. While that is somewhat an oversimplification, it describes what many successful mechanics do. To begin troubleshooting, consider complete *systems*, such as the ignition system, the cooling system, and so forth. Once you've identified the faulty system, begin eliminating *components* within that system. For example, if a fault has been traced to the ignition system, inspect the coil, distributor, plug wires and so on until the bad component is identified. The real challenge when troubleshooting is to *stay at the system level* and not jump prematurely to diagnosing individual components.

Consider the beginning mechanic trying to fix a rough idle. After dismantling the carburetor twice in an effort to find a problem with the idle circuit, this mechanic discovers the actual problem is a loose sparkplug wire. The engine was running on one less cylinder. The mistake here was to jump directly to one system that was presumed to be bad (carburetion) instead of first trying to verify that others were good.

Because this isn't a book on engine rebuilding, the details of troubleshooting mechanical problems, including lubrication, ignition, cooling, exhaust and suspension systems aren't included. Let's just say that the process of determining that something's wrong involves deduction; proceed from general to specific problem areas by elimination. If you don't deduce correctly, you end up tearing down a perfectly good carburetor when the ignition system needs attention.

System-Level Checks—Successful, efficient diagnosis begins at the largest possible categories, called *systems*. There are a number of systems within an automobile: chassis, electrical, ignition, carburetion, cooling, lubrication and mechanical are just a few. Systems are made up of components. The advantage of deductive testing is that you can ignore the condition of individual components if their system tests OK.

One reason inexperienced persons frequently confuse systems when trying to diagnose problems is that they aren't sure what the system is and what it does. An extreme case would be trying to stop an engine from smoking by replacing the muffler. Someone is confused about what each system does.

Typically, try to prove that an entire system is good to eliminate it from your diagnosis as quickly as possible. That is, if the chain is OK, assume each link should be OK, too. And there are some general system-level checks that can be made easily. If an engine doesn't make any abnormal sounds (whines, grinds and clicks, for example), you can probably conclude that its mechanical system (crankshaft, rods, pistons and so forth) is in good condition. If the engine doesn't overheat, presume that its cooling system is OK. If it shows good oil pressure, the lubrication system should be OK.

Traps—One of the common traps when troubleshooting is concluding that a system is good when, in fact, it isn't. Checking one sparkplug for spark is not a valid test of the whole system. This is why some engine analyzers are connected to the engine in several places to prove whether an ignition system is good or not.

If you don't know any reliable system-level checks, then don't try to invent them. You'll must check each component of the system before concluding that the whole system is good.

Another common trap is believing that a part is good because it has just been rebuilt or is brand new. Granted, it's a valid premise to assume that a part will work if new or rebuilt. But there is no guarantee the component is trouble-free and it's useful to test it (if possible) before installing it. For example, drop a thermostat in boiling water to make sure it opens.

Finally, there is the aggravating trap of the intermittent problem; the miss or stumble that happens just every once in a while. The real trap here is jumping to the conclusion that one system is causing the problem when another is the real culprit. It's easy to do because you never quite get a "handle" on the problem since it comes and goes without warning and usually when it's least convenient to check it.

A tactic in diagnosing an intermittent problem is to *make it repeatable*. Before you decide on a course of repair, you have to experiment until you cause the problem to happen. Then, make it happen again. Once you make the problem repeatable, you can begin to eliminate good systems. Note that making a problem repeatable does not immediately mean you've located the bad system. You should still work in an organized manner to verify good systems until only the bad one is left.

As a general guideline, the most difficult step in troubleshooting carburetor problems at the system level is distinguishing between ignition and carburetion as the source of the problem. The reason is that mechanical failures—a broken valve or rod, for instance—usually produce characteristic knocking noises. In contrast, electrical and carburetion failures usually happen without telltale noises, although, you may hear a loose or broken sparkplug wire arcing to ground, or intake air leaks whistling at carburetor and manifold surfaces and vacuum lines.

Moreover, both carburetion and ignition failures result in similar symptoms. Intermittent power loss, low power, failure to start, high-speed miss and rough idle are all symptoms that can be caused by either carburetion or ignition. As a result, you should be very careful before you conclude that a problem is electrical- or carburetor-related.

In general, electrical problems tend to produce symptoms that are very sudden—the engine shuts down all at once, with a jerk. Carburetion problems cause more gradual symptoms—the engine coughs, sputters a bit, and then dies.

There is one foolproof system-level check for carburetion: exhaust-gas analysis used with the car on a dynamometer. This requires equipment not found in your average garage. If you're not blessed with the necessary equipment, you'll have to do component-level checks to verify that the system is OK. Another alternative is to take your car to a garage that has such equipment and pay for the troubleshooting. Then you can correct any problems they identify.

WEBER DIAGNOSIS

At first glance, the carburetor, as a system, is composed of sub-systems that form the working circuits. The four circuits all Weber carburetors have are:
- Main
- Progression
- Idle
- Accelerator pump

Additional circuits most Weber carburetors have are:
- Starting
- Power

There are other sub-systems in Webers which are not circuits:
- Float
- Filter(s)
- Vacuum taps
- Solenoids

The rest of this book should help you understand which components are associated with each circuit and subsystem. What we do now, is to apply diagnostic logic to troubleshooting a Weber carburetor. Presuming that all the other systems (ignition in particular) and components have been proved good, only the carburetor is left, so it must be the source of the problem. What to check? The outline shows the logic:

SPECIFIC FAULT IDENTIFICATION
System-level Indicators
- Tailpipe deposits
- Sparkplug deposits
- Gas Mileage

Subsystem-level Circuit Identification
- Main Circuit
- Progression Circuit
- Idle Circuit
- Enrichment Circuits

FAULT IDENTIFICATION

So, before delving into the carburetor, check that the remainder of the fuel system is working. It may sound obvious, but first make sure there is gasoline in the tank! Fuel systems have been torn apart only because no one checked that there was fuel in the tank.

Take some precautions before working on the fuel system. Gasoline is extremely flammable, of course, so take steps to keep it from open flame or sparks. Do fuel system work in a well-ventilated location. Avoid breathing the fumes any more than necessary and wash fuel from your skin as soon a possible.

Don't check carburetor passages by

putting your mouth on them. Use low-pressure air blowing through windshield washer or vacuum line to check for a blocked passage.

Keep a *charged* fire extinguisher handy and know how to use it. Engine fires are quick and to the point—you don't need that experience.

Fuel Filter—There's usually a *fuel filter* somewhere in the fuel line. There may be a second, in the inlet housing where the fuel line attaches to the carburetor float bowl. Another *filter screen* may be located in the fuel tank. They don't usually clog, but can. Make sure that all filters are clean. If not, clean or replace them.

Fuel Pump—There's also a *fuel pump* between the fuel tank and carburetor. To test the pump, disconnect the fuel line at the pump's outlet and operate it, observing the volume of fuel that flows out of the pump. Collect the fuel in a non-breakable container.

The pump is either electrically or mechanically actuated. If it's electric, then turning on the ignition, but not starting the engine, should energize the pump. If it's mechanical, crank the engine, but don't start it. See the accompanying sidebar for tips on disabling the ignition.

The flow of gasoline from the pump should be forceful. If you have a pressure gauge, 3–5 psi is about right. Otherwise, the fuel should arc several inches as it exits the pump outlet. If it dribbles out, the pump is bad or the line or filter to the pump is clogged.

System-Level Indicators—Here are three system-level "eyeball" tests you can make which will give you an indication of the overall condition of the carburetion. These tests presume that everything else has been checked and found OK. Because they aren't done with instruments, they're only general tests, so some skill and experience are required to rely on them.

Tailpipe—You can make a tailpipe check on any car that doesn't have a catalytic converter or air pump. If your

---

DISABLING THE IGNITION

There are various methods to disable a conventional ignition system and short-circuit the electrical supply to the sparkplugs. The best way is to remove the high-tension lead from the center of the distributor cap and ground it.

This lead is the large, heavily insulated wire connecting the distributor cap's center to the coil. To remove it, turn the ignition off, grasp the boot around the distributor cap terminal, twist the lead slightly, and then pull it out. The twist helps break any corrosion that resists wire removal. Now, ground the lead to the head or engine block. Secure it.

On engines with *electronic ignitions,* there must never be more than a 1/4-in. gap between the free end of the lead and ground, or *ignition system damage may occur.* So, to ensure that there's no gap, use a jumper wire to ground the high-tension lead.

---

car has either, this test isn't accurate.

Look at the inside of the end of the tailpipe. It should be light brown, about the color of a chocolate malt. If it is, then the carburetor is probably all right. If the color is white, the carburetor may be set too lean. If black, run your finger around the inside of the pipe to feel the deposit. If it's oily, then the engine is burning oil and may require work. If the deposit is powdery, like charcoal, then the carb is probably set too rich.

If you can get a good tailpipe reading, it is the easiest way of fine-tuning the main jet for your style of driving. Just keep changing main jets until the tailpipe color is a chocolate-malt brown.

There's another quick test that involves the tailpipe. Start the engine and let it warm up, then blip the throttle hard and check for colored smoke from the exhaust. If the exhaust doesn't smoke, go on to another test. In general, if the exhaust smoke is blueish white—not to be confused with white steam—the engine is burning oil. If the exhaust is black, the carburetor is supplying an excessively rich mixture.

Be careful in deciding between black and blue smoke; it's not an easy decision, and this test requires some skill. If the smoke is definitely white, then it's probable coolant is leaking into the combustion chamber. You'll probably have to change a head gasket or fix a cracked block or head.

Sparkplugs—This test should be made in conjunction with with a tailpipe reading, because both depend on the color of the deposits. The sparkplug deposits you should examine are on the ceramic insulator that surrounds the center electrode. Use a magnifying glass.

To conduct this test, drive the car on an open road for a few miles at fairly high speed and then *cut clean.* This is done as follows. Without taking your foot off the accelerator, switch off the ignition (be careful *not to lock* the steering!) and immediately close the throttle, shift to Neutral and coast to a stop at a safe place. Now remove the sparkplugs, being careful not to burn yourself on the hot engine or sparkplugs.

The ceramic insulators of the plugs should be a light brown, just a bit lighter than the chocolate-malt brown of the tailpipe. If the insulators are white, the mixture is too lean. If the insulators are black, it's too rich. If the color of the insulators varies from cylinder to cylinder, there is unequal fuel distribution, probably because of poor manifold design. The manifold fault will have to be corrected before you can tune the carburetor with accurate results.

Mileage—This can indicate proper carburetor adjustment, though it is not as accurate a test as tailpipe or sparkplug insulator colors. Webers can yield better

33

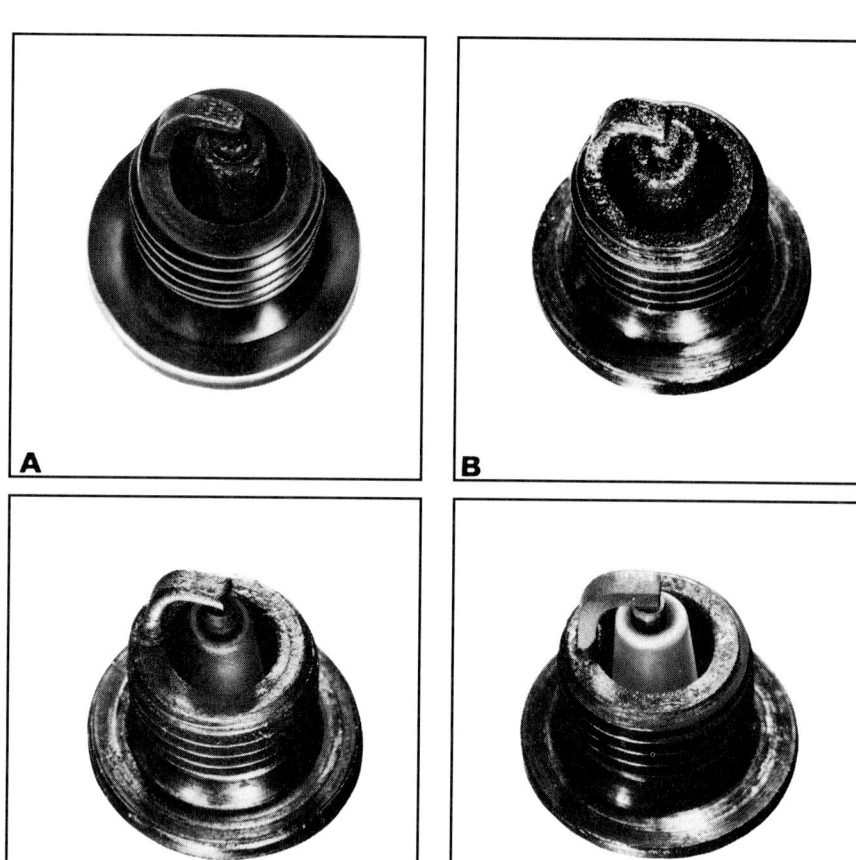

Reading sparkplugs can yield important troubleshooting clues. Plug A suffers from heavily rounded electrode and pitted insulator. It's worn out. Replace it. Plug B is oil-fouled. Shiny black coating indicates excess oil consumption, possibly from worn rings and valve guides. Plug C is carbon-fouled; don't confuse it with oil fouling. A carbon-fouled plug's dry, flat-black coating comes from excessively rich air/fuel mixtures, stop-and-go driving, or too-cold plug heat range. Plug D is normal. Electrode rounding is moderate and insulator is even tan or gray, indicating all is well in the combustion chamber. Photos courtesy Champion Spark Plug Company.

richens or leans all the circuits at the same time. On some Webers, the idle circuit is fed independently of the main circuit. Many tuners are satisfied just to get the main jet right. In fact, if you're not too critical, juggling main jet sizes will probably get you satisfactory driveability from a Weber.

Main & Air Jets—The main jet is almost exclusively responsible for fuel flow in the middle range of engine speeds. As engine speeds exceed mid-range rpm, the emulsion tube and air-correction jet become more significant. At maximum throttle, the air-correction jet has the most significant effect on fuel flow. It meters air to the emulsion tube. Thus, if the engine is running fine at, say, 2000 rpm, but rich at 6000 rpm, a larger air-correction jet is needed. Or, if the engine is running strong to 6000 rpm and then goes flat, the air-correction jet is too lean. Try a *smaller* air-correction jet.

Weber main and air-correction jets are stamped with their size in hundredths of a millimeter: *130* on a main jet means the jet flows fuel like an ideal jet that has a 1.30mm ID.

Emulsion Tube—The emulsion tube controls when the air-correction jet becomes significant, and the rate at which its significance increases. As a result, the emulsion tube affects driveability. Typically, no one experiments with emulsion tubes unless they have: 1) a very sensitive rapport with their car and a long time to experiment, or 2) a chassis dyno and HC/CO meter to detect changes in mixture caused by the emulsion tube.

For most novice tuners, being content with the original emulsion tube is the path of least resistance and best idea. Selecting the correct main and air-correction jets and emulsion tube is simple if you buy a *conversion kit* from a Weber supplier. Many will note your application and jet the Weber for you.

If you're making up your own conversion, and can find the jet values in an applications table, then jet selection is also easy. But, if you're really on your own and have no guidelines to follow, there are some general rules.

mileage than most other carburetors, all else being equal. This kind of test can require several months to perform properly, but you can use it to "dial-in" a carburetor setup to your own special driving style, if economy is your goal. Keep checking the tailpipe and sparkplugs to avoid going too lean or too rich as you change jets to improve mileage. Clearly, if your driving style is "pedal to the metal," then this test won't be of much interest.

CIRCUIT IDENTIFICATION

If everything else checks OK and you're convinced the carburetor is faulty, then isolate that part of the carburetor needing attention using the following guidelines.

MAIN CIRCUIT

The *main circuit* includes the *main jet, air-correction jet* and *emulsion tube.* Most of the fuel in the carburetor circuits flows through the main jet, so changing it

METRIC CONVERSION

Weber literature specifies dimensions in metric units. Consequently, if you're unaccustomed to relating engine displacement in cubic centimeters (cc) to cubic inch displacement (CID), here are some conversion factors.

Remember, 1000cc = 1 liter = 61 CID. Divide engine displacement specified in cc by 16.4 to displacement in CID. For example, a 2-liter engine equals 122 CID:
- 2 X 61 CID = 122 CID, or
- 2000cc ÷ 16.4 = 122 CID.

Going from CID to cc is also straightforward. A 350-CID engine displaces 5.7 liters (350 ÷ 61 = 5.7 liters = 5700cc. Or multiply CID by 16.4 to get displacement in cc. So, 350 X 16.4 = 5740cc; or simply 5.7 liters.

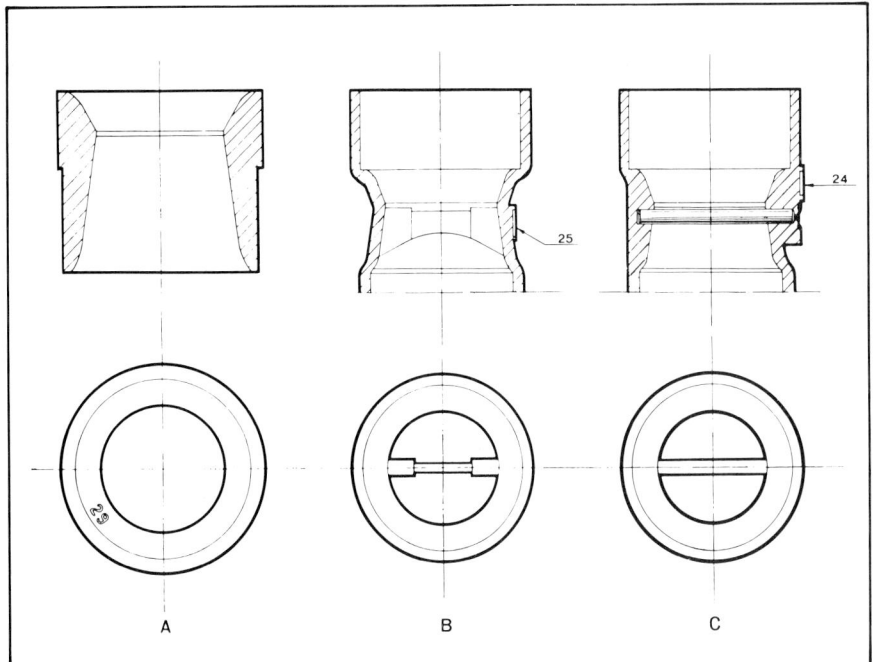

Main-venturi types: (A) is for DCOE having 29mm diameter, (B) is section of carburetor with internal non-removable main venturi of 25mm diameter with baffle for improved mixture distribution, and (C) is same type but with round bar serving same purpose as baffle. Drawing courtesy Weber.

Select Venturi—The first step in setting up the jets is to select the size of the *main venturi*, which determines the amount of air that will flow through the carburetor. The venturi size specified in millimeters (mm) is at its smallest inside diameter.

Venturi size selection depends on your application and objective for the engine. Is all-out horsepower at high rpm the goal? Then, in general, use a large-diameter venturi. If driveability at lower engine rpm is to be the primary use, then choose a smaller venturi.

Weber publishes a chart relating engine capacity to venturi size. It applies to 4-stroke engines with 1–6 cylinders and maximum output at about 5000 rpm, which are also fed by *one* single-barrel down- or side-draft carburetor. Following that chart, you'll see that a 2-liter 4-cylinder engine will require a 28–32mm main venturi. Thus, if you have a 2-liter, 4-cylinder Alfa, use a single-throat carburetor with a venturi of about 30mm.

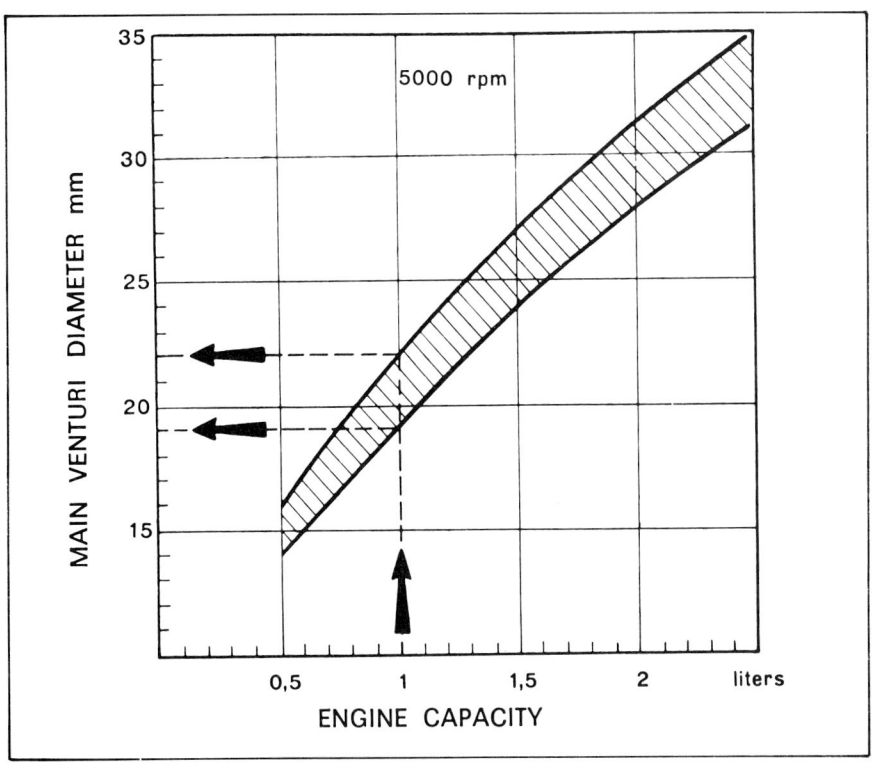

Weber chart specifies main-venturi selection (in mm) for four- and six-cylinder engines with maximum output at about 5000 rpm. Chart courtesy Weber.

35

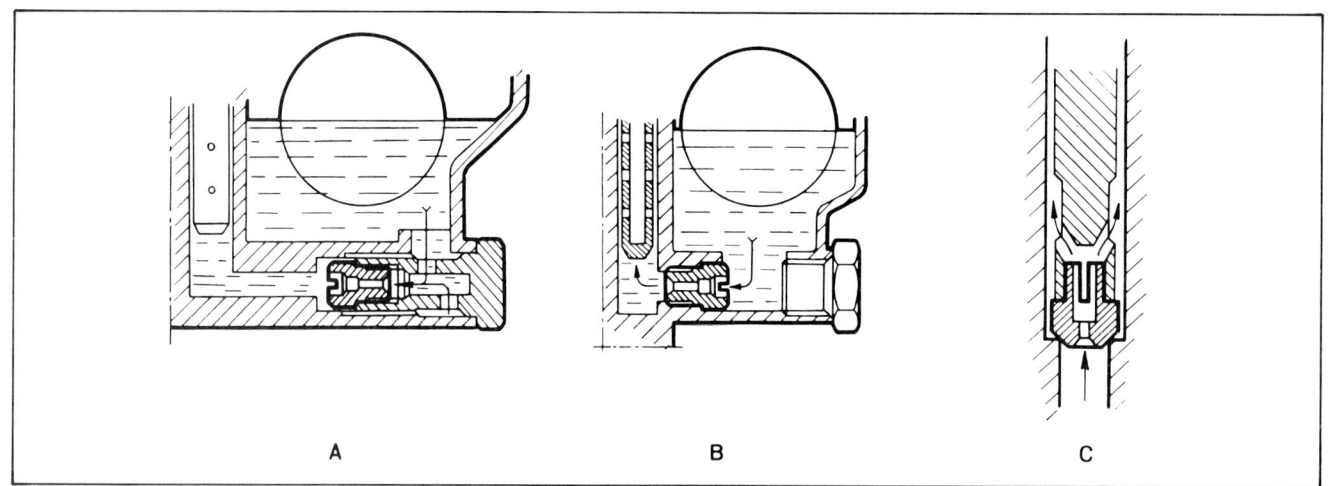

Main-jet locations: (A) jet installed in special holder, (B) screwed into carburetor casting, and (C) in-line with emulsion tube as in DCOEs. Drawing courtesy Weber.

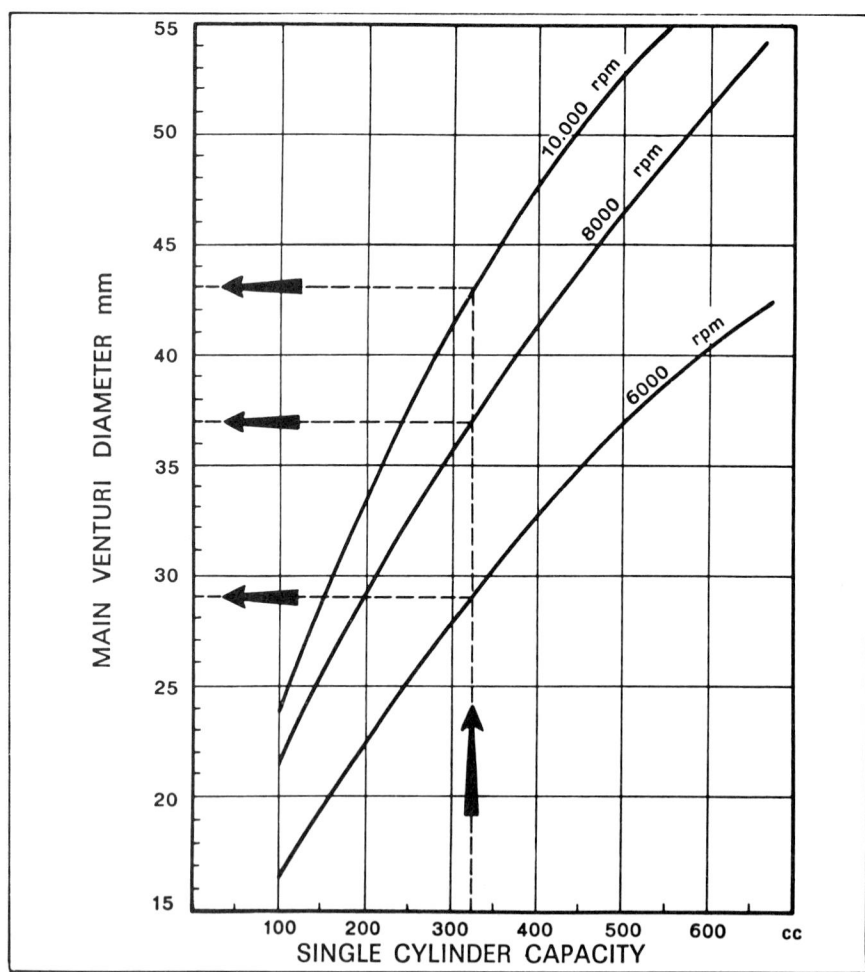

Weber chart specifies main-venturi size (in mm) for independent-runner manifold feeding single cylinder. Chart is calibrated for racing engines. Chart courtesy Weber.

Weber has another chart with general guidelines for venturi selection. It applies to 4-stroke engines with one carburetor barrel per cylinder. This chart specifies venturi size for independent-runner manifolds and single-cylinder capacities.

Using the 2-liter Alfa as an example again, note that its single 500cc cylinder (2000cc divided by 4 cylinders = 500cc) needs a 37mm venturi to breathe sufficiently at 6000 rpm according to the chart. Seems a bit large compared with the 30mm venturi specified by the first chart. A check of the Weber applications list shows that a 32mm venturi is actually used in the Alfa's 40 DCOE Webers. This Weber chart for independent-runner manifolds and one carburetor throat per cylinder is for *very* sporting engines. Indeed, all-out race engines.

Weber has developed two charts that cover the extremes: a single-barrel carburetor for somewhat mildly tuned production engines, and another for barnburners that can turn as fast as 10,000 rpm. The solution is to pick something in between, which is exactly what the Alfa's 32mm venturi for 2-liter displacement represents.

Select Main Jet—Having selected a venturi size, now select the main jet. Garry Polled of TWM Induction uses a general rule: When selecting the main jet for a DCOE, multiply the venturi size by four to get an approximate *beginning* main jet size. He bases this on his 20-plus years of experience with Webers. In our example:

32mm venturi × 4 = 128 main jet

A check of the application chart shows the Alfa running 135 main jets in its 40 DCOEs. That's fairly close to the rule, though a 128 jet would probably be a bit lean.

To select the air correction jet for this DCOE, Garry has another general rule: Add 60 to the main jet size. Continuing with our example Alfa:

128 main jet + 60 = 188 air-correction jet

In fact, the Alfa runs a 210 air correction jet. But, because it also carries a main jet a bit richer than our rules of thumb indicate, the larger air-correction jet makes sense.

Clearly, Polled's guidelines are starting points, as the actual application example with the Alfa shows. But, if you have nowhere else to start when tuning a DCOE, here's the rule again:

Main jet = venturi diameter × 4
Air correction jet = main jet + 60.

Drop all decimal values. In fact, the venturi diameter is in millimeters, while actual jet diameters are in hundredths of a millimeter. For the rule of thumb, the decimal point is ignored. Again, this is a starting, not an ending point in jetting a Weber.

EMULSION TUBES

Emulsion tubes are used in both the main and progression circuits of Weber carburetors. Their purpose is to "bubble" or emulsify the fuel, slowing it and making it flow closer to the flow rate of air.

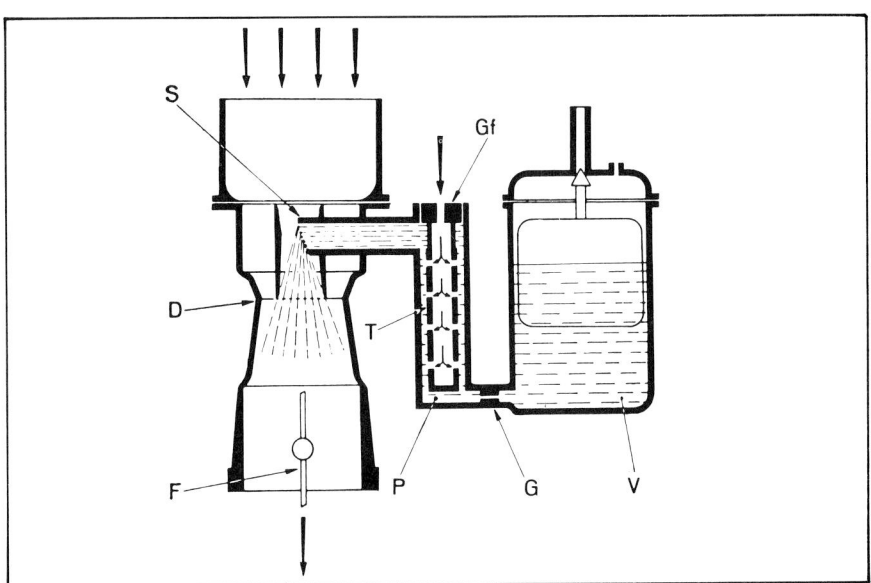

Emulsion tube (T) fits into well of fuel (P) and bubbles air into fuel to slow its passage. Other parts here: (S) spray nozzle, (Gf) air-correction jet, (G) main fuel jet, (V) float chamber, (D) venturi, and (F) throttle. Drawing courtesy Weber.

Without emulsion tubes, a carburetor would run progressively richer as flow rates increase.

Weber is somewhat terse in its description of the way an emulsion tube operates. Combine that terseness with the chaotic emulsion tube identification system, and you have all the elements of a good mystery. Not surprisingly, then, emulsion tubes are a mystery to many. And the effect of changing emulsion tubes is very subtle to detect. So the results of a change can be hard to judge, and often impossible to determine if the change is small.

Weber regards the emulsion tube as a *brake*. That is, the purpose of the tube is to slow down the passage of gasoline, just as a brake retards the motion of a car. The tube braking action is regulated by the:

● Outside diameter of tube.
● Number of holes in tube.
● Orientation of holes in tube.
● Volume of air passing into tube (controlled by the air-correction jet).

Let's examine each of these parameters in turn:

Diameter—The emulsion tube fits into the *well*, a column of fuel supplied through the main jet. The fuel level of the well is as deep as the float level of the main fuel bowl. If there were no emulsion tube or air-correction jet, the well would be nothing more than an auxiliary reservoir of fuel.

The outside diameter of the emulsion tube, as it fits in the well, acts as a barrier to the passage of fuel. If the emulsion tube fits tightly inside the well, no fuel at all would pass.

Some emulsion tubes have a constant outside diameter. For them, the size of the outside diameter limits the amount of fuel that can be drawn past the tube. The *larger* the outside diameter of the tube—the less clearance between tube and well—the *less* fuel that can be drawn past, and the *leaner* the mixture.

Some emulsion tubes are machined with a *step*, so that the major (largest) diameter occurs *below* the first level of holes. The location of this step controls the pressure at which the venturi begins drawing fuel from the well; therefore, the point at which the main circuit engages.

The *lower* the step, the *earlier* the main circuit starts and vice versa. The significance of the step depends on the

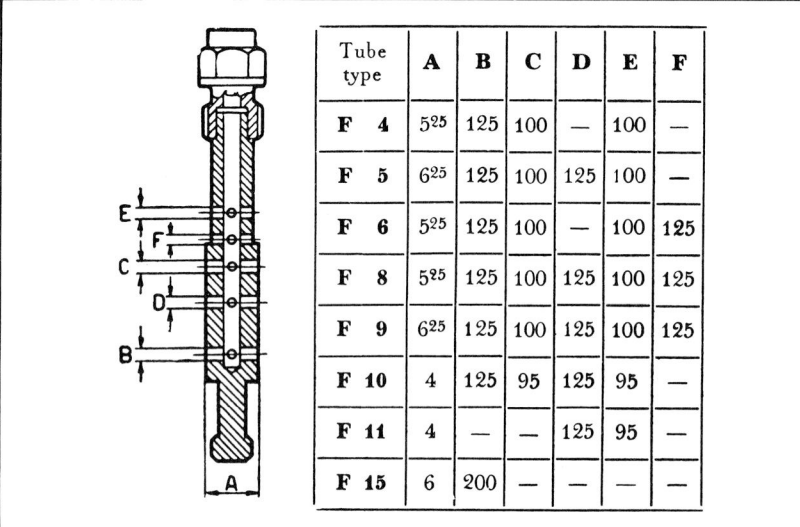

Tube type	A	B	C	D	E	F
F 4	5^{25}	125	100	—	100	—
F 5	6^{25}	125	100	125	100	—
F 6	5^{25}	125	100	—	100	125
F 8	5^{25}	125	100	125	100	125
F 9	6^{25}	125	100	125	100	125
F 10	4	125	95	125	95	—
F 11	4	—	—	125	95	—
F 15	6	200	—	—	—	—

Note that distance between layers of holes decreases from bottom to top. Spacing helps adjust fuel flow depending on level of fuel in well. At high speeds, when fuel level in well is lowest, there is greatest braking action. The diameter and position of top row of holes are most critical, for they determine, along with diameter of air-correction jet, point at which braking action begins. Drawing courtesy Weber.

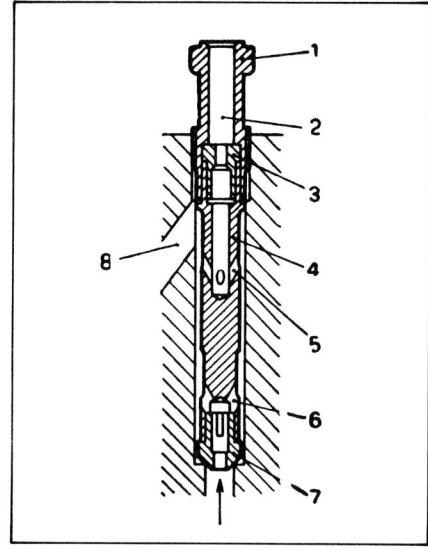

F16 emulsion tube has upward facing air-correction holes (5) and fuel holes (6). Another "family" of Weber emulsion tubes is represented by F16 tube. Its holes angle up, helping, rather than retarding, fuel flow. As a result, F16 emulsion tube causes main circuit to start earlier than F4—F15 "family." Drawing courtesy Weber.

float-bowl level. And that's one reason why setting the float level correctly is important. A higher float-bowl level has the same effect as lowering the step, and a lower float-bowl level has the same effect as raising the step. The step ensures that, even at the lowest speeds, there will be a bubbling effect on fuel drawn past the emulsion tube.

Just as in the case of the constant-diameter emulsion tube, as the major diameter of the tube decreases, less signal from the venturi is needed to pull fuel past the tube. *Signal* means the amount of pressure drop at the venturi, which "signals" the fuel-supply circuits. Thus, the smaller the outer diameter of the emulsion tube, the earlier the main circuit starts and the less the braking action of the emulsion tube.

Number of Holes—Air is forced through the holes in the emulsion tube as it tries to reach the low-pressure area of the venturi. The air must bubble through the fuel in the well to reach the venturi. If there were only one or two small holes in the emulsion tube, very little air would be able to pass from the well to the venturi. Consequently, the more holes and the larger they are, the greater the braking action on the fuel, and the leaner the mixture.

The holes at the top of the emulsion tube bubble fuel at low speeds. As the velocity of the flow increases, fuel is forced farther down in the well by the increasing flow of air through the emulsion tube. Thus, the spacing of the holes down the tube, as well as the number and size of the holes, controls the overall air correction of the fuel flow.

Typically, the holes in the emulsion tube are more closely spaced at the top, where working pressures are lower and precise control more critical. Holes at the bottom of the emulsion tube operate at very high pressures and have the most profound effect on the system.

Hole Orientation—The typical emulsion tube has holes drilled at right angles to its axis. This "family" of tubes is specified in the accompanying chart.

Note that the distance between the layers of holes decreases from bottom to top. Hole spacing helps adjust fuel flow depending on the level of the fuel in the well. At high speeds, when the fuel level in the well is lowest, there is greatest braking action. The diameter and position of the top row of holes are most critical, for they determine, along with the air-correction jet diameter, the point at which braking action begins.

Another "family" of Weber emulsion tubes is represented by the F16 tube, illustrated here. Its holes are angled upward, helping, rather than retarding, the flow of fuel. As a result, the F16 emulsion tube causes the main circuit to start earlier than the F4–F15 "family."

Air Volume—The amount of air available to the emulsion tube is controlled by the air-correction jet. In some Weber applications, the air-correction jet is a separate part from the emulsion tube. In those applications, the emulsion tube is held in the well by the air-correction jet. In other applications, including the idle-jet assembly, the air-correction jet is part of the emulsion tube.

In the DCOE carburetor, and most idle-jet assemblies, the air-correction jet, emulsion tube and main jet are a single push-together assembly that can be changed very easily.

The size of the air-correction jet has an inverse relation to the air/fuel ratio—the larger the jet, the less fuel; the smaller the jet, the more fuel passed to the venturi.

PROGRESSION CIRCUIT

The progression circuit has the most profound effect on driveability, and is the circuit most used in everyday driving. It's the circuit that allows the carburetor to maintain the same mixture strength as it makes the transition between the idle and main circuits. Engines that stumble as they pull away from a stop or turn a corner slowly probably have a progression circuit needing attention.

The progression circuit receives fuel through the same kind of jet/air-correction circuit that feeds the main circuit, so much of the discussion regarding the main circuit and its air correction also applies to the progression circuit. In overall terms, fitting a larger idle jet will richen the progression circuit; fitting a smaller idle jet will lean the progression circuit.

When there is some question that the progression circuit is too rich or too lean, there's a natural tendency to try to overcome its fault simply by readjusting the *idle mixture screw*. Remember that the progression circuit comes into play as soon as the throttle plate moves. So, changing the idle mixture richness screw won't affect the operation of the progression circuit. Thus, it's important to *change* the idle jet and not adjust the idle mixture screw to troubleshoot the progression circuit.

Depending on the type of carburetor, you can also select different emulsion tubes for some progression circuits. The differences the emulsion tubes make is very, very subtle. It's better to stick with the tube fitted and adjust the idle jet to get the desired mixture.

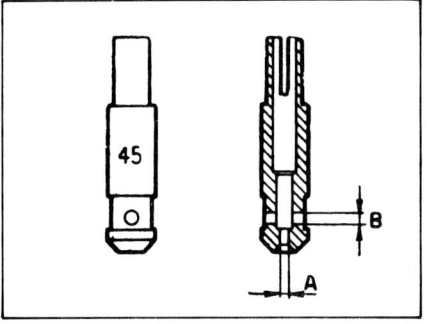

Idle jet with calibrated fuel circuit: (A) is idle-jet diameter, (B) is uncalibrated fuel passage. Drawing courtesy Weber.

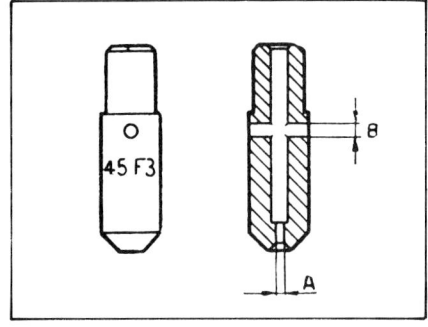

Idle jet with calibrated fuel- and air-correction circuits. (A) is idle-jet diameter, (B) is calibrated fuel passage. Drawing courtesy Weber.

Idle jets—There are two kinds of idle jets for Weber carburetors: those with both fuel- and air-correction calibration, and those with only the fuel jet calibrated. Idle jets incorporating a letter, for instance *45 F3*, have both air and fuel flow calibrations in the jet. Jets with only a number, for instance *45*, only control fuel flow. Air-correction calibration is then located somewhere else in the carburetor itself.

Testing—The test for progression circuit operation is accurate, but requires some time, patience and attention. With a warm engine, set idle as slow as the engine will continue to run. Very slowly increase idle speed using the idle-speed adjustment screw. Stop increasing idle speed at about 2000 rpm, or whenever you're sure the main circuit has begun to take over fuel delivery—you may be able to hear it come in.

At some point in this slow process, the throttle plate will begin to uncover the progression circuit openings in the throttle bore. After this point, the progression circuit will determine air/fuel mixture until the main circuit begins to come in.

Actual engine speed at which the main circuit comes in will vary, of course, but it should definitely be in operation by 500 rpm above minimum idle; 1300 rpm is a good rule-of-thumb figure. If everything is set correctly, there won't be any noticeable change in smoothness of the engine during the slow idle-up process.

If the progression circuit jetting is incorrect, there will be a point at which the engine begins to run rough or surge, or the exhaust note will change and probably be somewhat fluffy and irregular. As engine speed increases, these symptoms will diminish as the main circuit takes over.

If engine smoothness changed, repeat the slow idle-up procedure until you are sure the progression circuit is in operation. Back out the idle mixture richness adjusting screw 1/2 turn. If the engine runs more smoothly, the idle jet is too small (lean) and should be larger.

If the engine runs rougher, then the idle jet is too large (rich), and should be smaller. If there is no change in engine roughness, turn the idle mixture screw in 1/2 turn to its original position, then in another 1/2 turn. If the engine smooths out, the idle jet was too large and should be smaller. If the engine runs rougher, then the idle jet is too small and should be larger.

Return the idle mixture screw to its original position, replace the idle jet as necessary and repeat the test. Changing the idle jet and re-running the idle-up procedure should correct the problem. If there's a separate emulsion tube for the idle circuit, *resist* the urge to change it, unless you're equipped with a chassis dyno and emissions testing equipment.

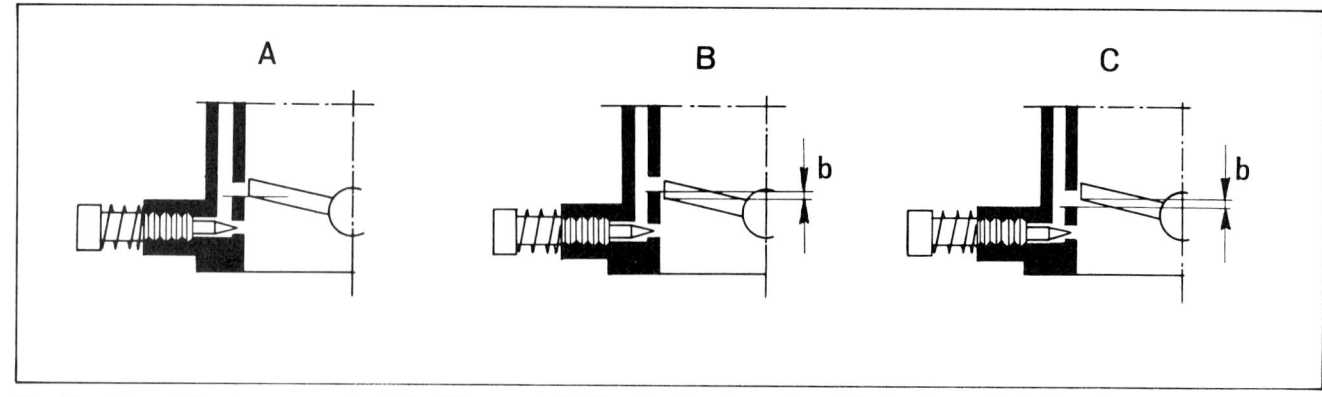

Throttle plate relation to progression hole: (A) correct; (B) throttle plate is too low; (C) plate is too high. Drawing courtesy Weber.

Throttle Plate Location—Some final comments about the progression circuit. All the above is based on the assumption that the throttle plate is correctly located. If the Weber is box-stock, that's a safe assumption. If you or anybody else have had the throttle plates out, however, it's possible that they are not *exactly* where they should be.

Two possible misalignments are shown in the accompanying illustration. Ideally, the progression hole(s) should be uncovered just as the throttle plate begins to move. But in actual practice, the plate may be set too low or high in the bore.

Plate Too Low—If the throttle plate is too low, it must move too far to activate the progression circuit, and there will be an excessively lean point (*flat spot*) in acceleration which no amount of fiddling will entirely solve. Try to reposition the throttle plate to correct this.

This isn't a casual task. The throttle plate screws are *staked* (the ends are deformed) to keep them from vibrating loose and making their way into the engine. They may break off or their threads may strip when you try to loosen them. See page 50 for more details on loosening, removing and restaking the screws and throttle plate alignment.

As a last resort, to correct a too-low throttle plate, the lower edge of the throttle plate can be chamfered with a file to

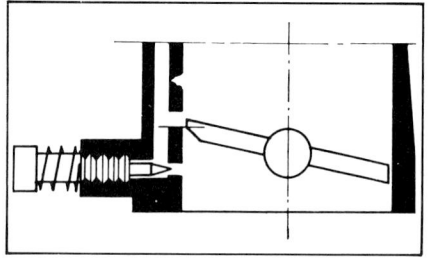

Cure for late-opening progression hole—example B in above drawing: chamfer lower edge of throttle plate. Drawing courtesy Weber.

uncover the progression hole more quickly. *Don't,* under any circumstance, try to bend the throttle plate.

Plate Too High—If the throttle plate is too high, it will always leave some of the progression hole uncovered even when the plate is fully closed. In this case, the idle mixture will be too rich. If repositioning the throttle plate won't solve the condition, a *very* small hole, no larger than 1.5mm (0.06 in.) can be drilled in the throttle plate on the side *opposite* the progression hole. This hole will allow additional air to flow into the engine and correct the over-rich idle mixture. The hole should be enlarged very carefully, in steps, to obtain the right correction.

Only drill the throttle plate if all other tuning efforts fail.

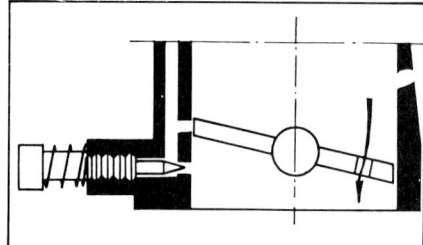

Correction for high throttle plate—example C in above drawing: drill *very* small hole, no larger than 1.5mm/0.06 in., and carefully enlarge it to obtain correct idle mixture. Drawing courtesy Weber.

IDLE CIRCUIT

The idle circuit is easy to tune, because there's an adjustable needle jet (screw) in it, placed near the base of the carburetor. With the engine at operating temperature, set the idle speed to about 700 rpm. Then slowly back out the idle mixture richness screw until the engine speed starts to drop off. Slowly turn in the screw until the engine has reached its highest idle speed. That's all there is to it.

Older carburetors have an idle mixture richness screw with a fairly sharp taper—1/2 turn makes a significant difference in idle quality. Modern idle mixture screws have a more shallow taper (are more pointed), so whole turns are required to make a perceptible change in mixture richness.

Very small engines using Webers

occasionally won't idle at all. At anything over idle, they run fine, but will die as soon as they drop under about 1500 rpm. The idle jet size on some Webers is so small that the fuel filter won't block dirt that can clog the idle jet. Remove the idle jet and blow through it with compressed air to clear the dirt. Never poke into a jet with a wire; you may scratch it and upset the precise calibration.

Some engines won't respond to idle adjustment. No matter how you adjust the idle mixture richness screw, there's no perceptible change in engine speed. Some possible causes include: 1) the carburetor is either emission-controlled (idle mixture changes very little over the available range of adjustment), or 2) fuel is entering the manifold from another circuit, most probably the progression circuits, as a result of too-high idle speed, or 3) the idle circuit itself is blocked.

ENRICHMENT CIRCUITS

Choke—The choke is the most obvious enrichment circuit, and when it doesn't work the engine won't start in cold weather unless you pump the accelerator many times. Some Webers use a *butterfly* choke; others have a separate circuit that acts as an auxiliary carburetor.

Butterfly Choke—This setup can be cantankerous because, being adjustable, more often than not they are mis-adjusted by the average tuner. My advice for most of us is: If it ain't broke, don't adjust it.

In theory, the butterfly is lightly spring-loaded so it closes completely when the engine is off. As soon as the engine starts, and the accelerator pedal is depressed, a fast-idle cam springs into place to keep idle speed up, and the rush of air into the carburetor opens the butterfly slightly. As the choke mechanism is warmed—either by its own electrical heater, which is fast, or by engine coolant, which is slower—the butterfly is opened by a bi-metallic coil. Spring force from the bi-metallic also serves to rotate the fast-idle cam so its effect is reduced.

How do you check this complicated

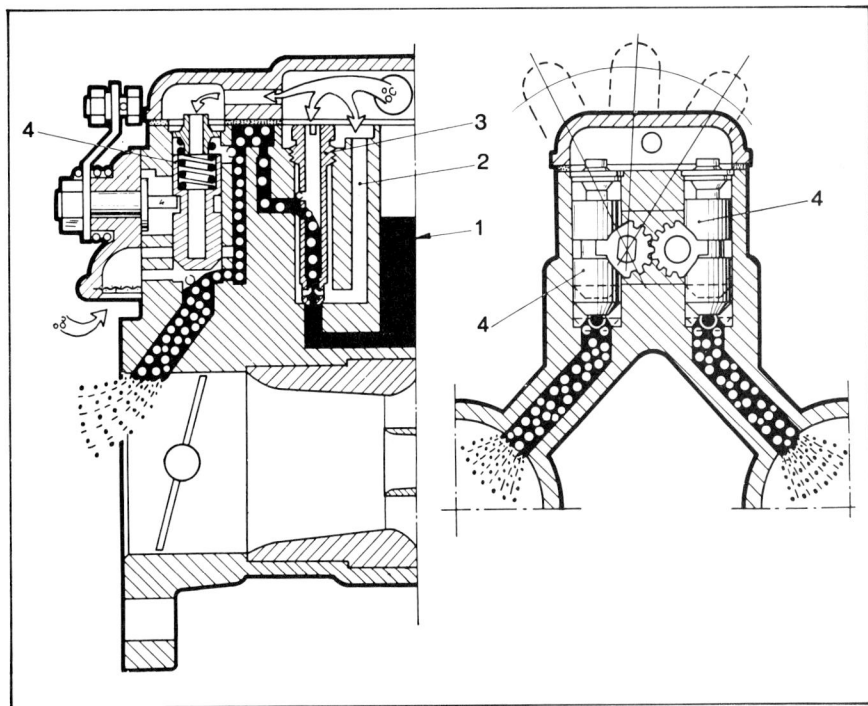

DCOE choke schematic: (1) float bowl, (2) starting fuel well, (3) choke jet with integral emulsion tube and air jet, and (4) plunger valve. Drawing courtesy Weber.

operation? A coarse method is to remove the air cleaner when the engine is stone cold (car has sat overnight). The butterfly should be closed. Start the engine, let it warm up for about three minutes and look at the butterfly again. It should be all the way open. If it isn't, check the action of the actuating rod. It may be gummed up and cleanable with standard carburetor cleaner. Any correction beyond that usually requires getting help from a Weber specialist.

Another alternate is to disable the butterfly by disconnecting its operating linkage and wiring it so it stays open. You'll have to change your starting procedure, though, and withstand poor initial driveability—particularly in cold climates.

Prime the carburetor by pumping the accelerator pedal 3–5 times. Then start the engine and hold the accelerator pedal at a fast idle until the engine warms up enough to idle on its own. This isn't an elegant solution, but will keep the engine running until you tackle the choke mechanism and correct it.

Auxiliary Venturi Choke—This choke type doesn't use a butterfly plate to richen the cold-start mixture. It is used most notably in DCOE series carburetors and is manually operated via a cable. The cable attaches to a lever on the carburetor which controls the raising and lowering of two (one for each barrel) *plunger valves*. These valves control the flow of extra fuel from the float bowl. The extra mixture enters the carburetor barrel on the engine side of the throttle plate. When the plunger valves are fully raised, extra mixture is allowed to pass (choke on); when they are fully seated, mixture flow stops (choke off).

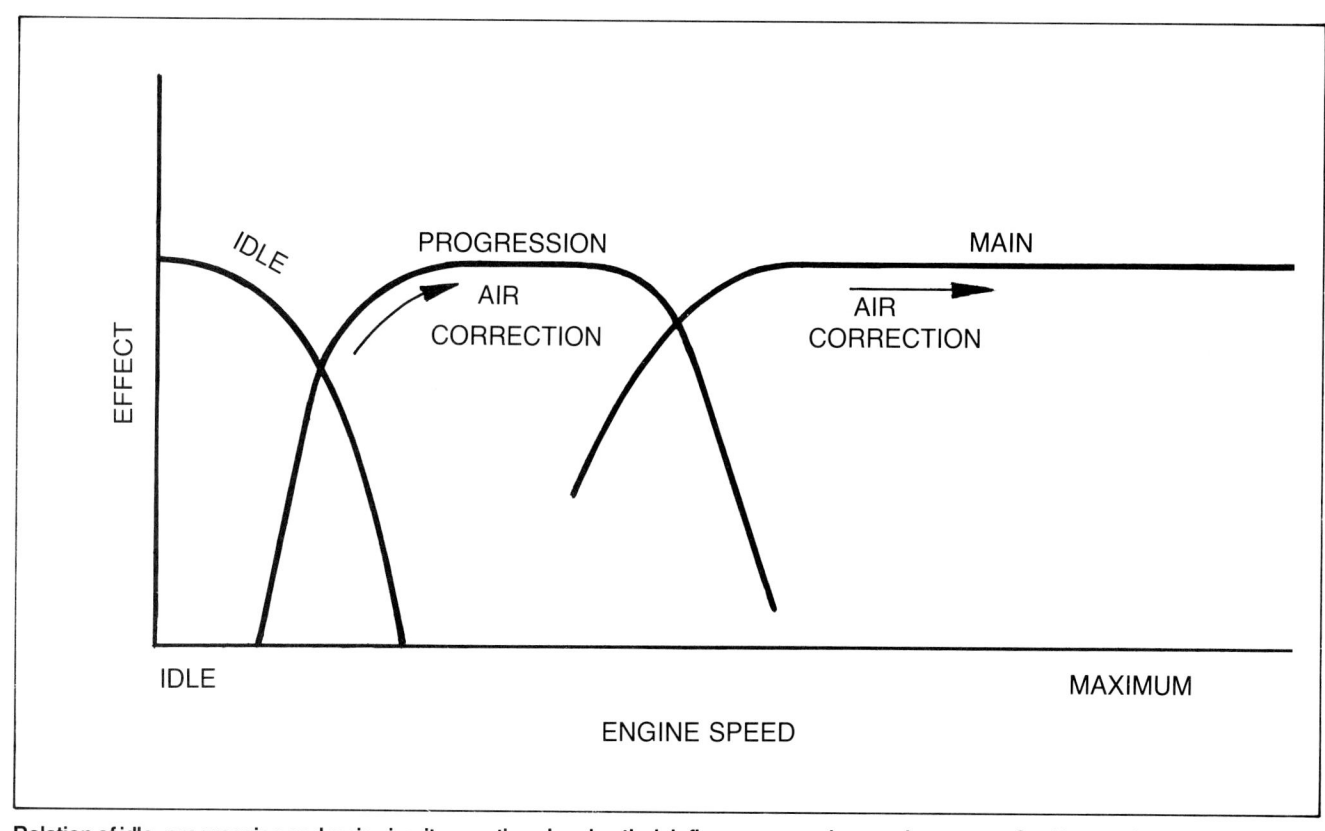

Relation of idle, progression and main circuit operation showing their influence as engine rpm increases. Goal is to make mixture strength a straight (stoichiometric) line. Actual practice is less than ideal.

One problem with this straightforward choke is that the cable operating the choke lever may keep the choke on even when you think it's closed. These types of chokes make a definite hissing noise when they're operating. Disconnect the actuating lever from the operating cable and move the lever by hand. Note where the lever is as you hear hissing from the choke. Reconnect and adjust the wire to the lever so the choke operates correctly.

Accelerator Pump—The accelerator pump is another fuel-richening device. Fortunately, it's easy to check. With the engine off, peer down the primary venturi—primary venturi on a progressive carburetor—and operate the throttle linkage. You should see a solid squirt of fuel coming from the pump jet nozzle. If you don't, the pump isn't working. On DCOE carburetors, the outlet is very close to the throttle plate. On most other Webers, it's near the top of the carburetor throat.

Power Valve—Modern engines are designed to run lean in order to pass exhaust emission standards. To maintain mixture strength, a *power valve* is used. It dumps a little extra fuel into the fuel system when manifold pressure falls below a certain value. The power valve is usually contained inside the carburetor and cannot be checked without disassembling the entire unit. If the engine has become sluggish while accelerating at higher speeds, or runs *very* rich all the time, suspect the power valve.

Fuel Level—A carburetor works because of the difference between atmospheric pressure on the fuel in the float bowl and the reduced pressure created by a restriction in airflow. The weight of the fuel in the well between the float bowl and venturi must be subtracted from atmospheric pressure in calculating the actual pressure developed. It's not a big value, admittedly, but it is significant. The weight of the fuel in the well is actually controlled by the float level of the carburetor. The higher the level, the more fuel in the passage.

As a result, float level changes mixture richness. Moral: Check the float level. Float level can be set so high that fuel actually overflows into the throttle bore, even at idle.

REPAIR METHODS

You should now be able to identify most problems afflicting a Weber, either by deduction, or by a direct hint from the practical diagnostics section. So after identifying a carburetor problem, proceed to correct it. The decision process can be broken into these courses:

- Non-catastrophic problem—correct by adjustment.
- Catastrophic problem—correct by replacement.

This boils down to, "Can I fix it, or should I replace it?"

If the problem is just a torn or worn-out

gasket, then the significance of a such an inexpensive part isn't great. But, if you are considering replacing the entire unit and facing that expense, then try to fix it before spending so much. Most Weber repairs are between these extremes, but the decision is the same—fix or replace?

Webers are fabulously fixable because they have so many replaceable parts: all major jets, venturi, diaphragms and so forth. About the only catastrophic failure you'll find is when the main casting itself is damaged.

So what do you do if faced with a total Weber rebuild? First, you must use all the special tools to reform jet seats and verify the diameters of internal passages. I recommend that you pass on that experience; it could be costly.

The reason major overhaul is best avoided is that the time involved probably exceeds the value of replacing the defective unit with a new one. And the experience required is beyond the level of most of us.

So, the action that represents the best compromise of cost, time and results is to do a thorough cleaning and to install a set of new gaskets, or a tune-up kit. A *gasket kit* includes all gaskets, seals and washers for the particular model. A *tune-up kit* has a gasket set, float, float fulcrum, needle valve, mixture screw(s) and fuel filter. There may be other parts included which apply to other applications.

Overhaul kits are available, too. They have all the pieces of the tune-up kit plus an oversize throttle shaft (applies to units without ball bearing), ball bearings, throttle plates and screws. Check the Suppliers List at the end of the book for places that sell these different kits.

With these different levels of kits available, you should be able to correct most carburetion problems caused by worn parts. Nevertheless, there are some repair procedures you should not perform when attempting to correct a problem with a Weber.

Leaking Float—Some floats are made of two thin brass stampings soldered together so they are airtight. Such floats can develop leaks, fill with gasoline, and sink. The entire metering action is upset—essentially the float chamber is registering empty when it isn't. A sunken float opens the needle valve, overfills the fuel bowl and causes an excessively over-rich mixture.

Many modern floats are made of a closed cellular material that doesn't leak. Sometimes they slowly absorb gasoline and lose their buoyancy. Again, their metering action will be skewed. You can weigh these floats and match their actual weight to the manufacturers specifications. For example, a 40 DCOE float should weigh 26 grams. (Brass floats can be checked this way too.)

One last suggestion: When setting float level, recheck it. It is critical for the carburetor's proper operation.

The carburetor I've sketched so far will work fine for an engine running at constant rpm. Indeed, the earliest engines operated at such a low speed that they were almost constant-speed engines. As engine rpm rises, though, the need for additional fuel and air must also rise to maintain stoichiometery.

There is a significant danger of an explosion while trying to resolder a metal float. Gasoline fumes will certainly be trapped inside a float that has leaked. It's unlikely that the float can be heated well enough to run new solder continuously all the way around it without using an open flame. **Don't attempt to repair a leaking float.** Replace it!

Drilling Jets—Weber jets are marked *after* they are drilled. The value marked on the jet is a measured value determined at the flow bench. When you drill a jet to try for a larger size, you have lost a basis for calibration; the size of the hole is only one of the factors determining the jet's ability to flow fuel. Hole finish and entrance and exit shape are other factors affecting flow.

Don't store jets by stringing them on a wire. Their calibration can be upset if the wire scratches them. Instead, put them in envelopes or separate containers. A small tackle box has plenty of storage for a selection of jets.

Cleaning Passages—Don't clean the orifices in the carburetor with wire. First, you could break the wire in a passage and not be able to extract it. That will be an expensive lesson you'll not soon forget. Second, the wire may gouge the precision passages and upset fluid flow. You can *carefully* drill out plugs and blow passages free with compressed air (a tire pump works fine in many cases), but avoid using metal probes in the passages.

5
Repair

TWM Induction installation of six IDFs on Jaguar V12: Too bad Garry Polled's sanitary work is hidden when air cleaner is in place.

WEBER REPAIR

In this chapter, five carburetors were chosen to represent all Weber carburetors; obviously, something has been left out.

What's missing are the nitty-gritty individual differences of the Weber product line. Each carburetor illustrated in this section is almost identical to many others; indeed, whole classes of carburetors are represented by a single model. The alternative to this approach would be a very thick volume. Perhaps it would make a nice reference, but would be unreadable. It would also be unnecessary. Once you understand the design fundamentals, you should be able to charge into most Webers with confidence.

In fact, there is a *Weber way* of building a carburetor. Once you are familiar with the basic techniques Weber uses, all Weber carburetors are an "open book." The purpose of this chapter is to present these design fundamentals, at the cost of glossing over individual details.

To understand how one of the five carburetors covered in this section relates to the Weber you want to repair, review the accompanying chart. The five carburetors covered in this repair section are shown in bold type.

There are two major divisions of Weber carburetors: horizontal (sidedraft) and vertical (downdraft) throats.

Sidedraft—The DCOE carburetor was selected to represent the horizontal-throat or *sidedraft* group. The single-throat OC is standard equipment on the Type 207A Abarth, if you happen to have one of the four known examples. All is not simple with Weber sidedrafts. There is perhaps as much variety within the DCOE range as for the entire remainder of the chart. Throttle bore sizes range from 38—58mm, and the DCOE is standard equipment on cars such as Alfa Romeo, Aston Martin and Maserati. It's also the carburetor of choice for most specialty high-performance, in-line engines, as well as some modified V8s.

Downdraft—Four carburetors have been selected to represent the range of vertical-throat or *downdraft* Webers.

The IMPE was selected from the group of single-throat Webers because it is both old (standard on some Fiat 600s) and simple (not a single emission control). When overhauling a modern single-throat ICH, you will find a different choke mechanism (butterfly valve vs. the IMPE's separate fuel circuit), a dia-

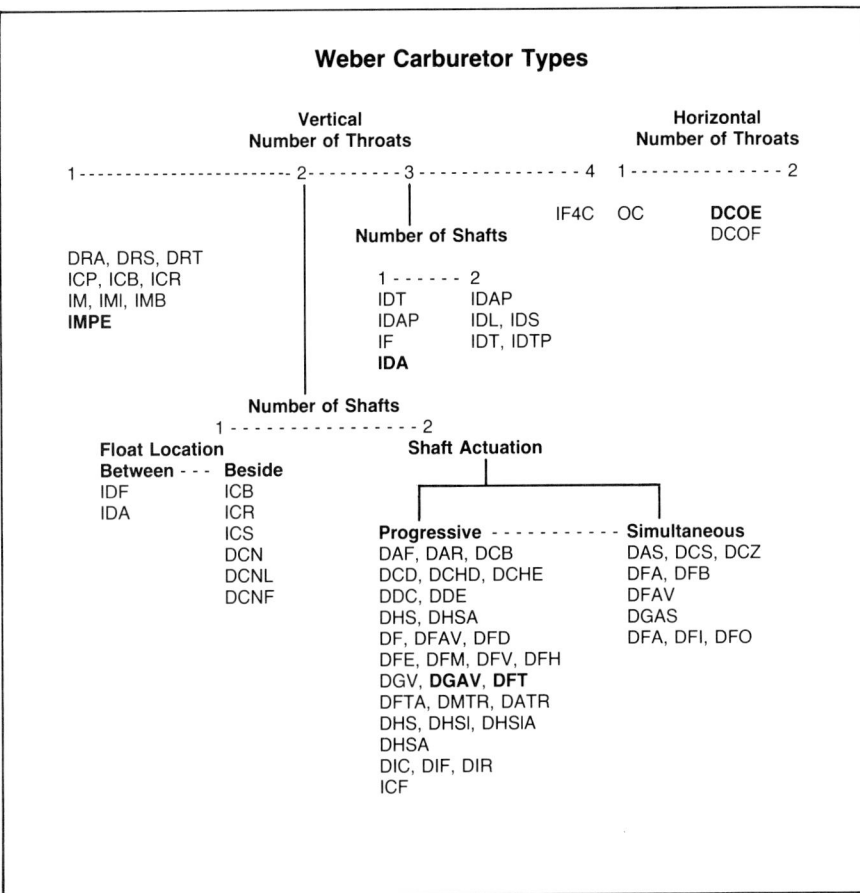

Weber Carburetor Types

```
                    Vertical                              Horizontal
                Number of Throats                      Number of Throats
1 -------------- 2 --------- 3 --------------- 4     1 -------------- 2
                             |                IF4C  OC       DCOE
                     Number of Shafts                        DCOF
                        1 ------ 2
DRA, DRS, DRT          IDT      IDAP
ICP, ICB, ICR          IDAP     IDL, IDS
IM, IMI, IMB           IF       IDT, IDTP
IMPE                   IDA
                     Number of Shafts
                        1 ---------------- 2
Float Location                    Shaft Actuation
Between --- Beside
IDF         ICB
IDA         ICR
            ICS            Progressive ---------- Simultaneous
            DCN            DAF, DAR, DCB          DAS, DCS, DCZ
            DCNL           DCD, DCHD, DCHE        DFA, DFB
            DCNF           DDC, DDE               DFAV
                           DHS, DHSA              DGAS
                           DF, DFAV, DFD          DFA, DFI, DFO
                           DFE, DFM, DFV, DFH
                           DGV, DGAV, DFT
                           DFTA, DMTR, DATR
                           DHS, DHSI, DHSIA
                           DHSA
                           DIC, DIF, DIR
                           ICF
```

IMPE has relatively few parts, but is well-made. *Sliding piston* starting circuit and piston-type accelerator pump is similar to sporting carbs like DCOE.

WEBER SPECIAL TOOLS

Weber makes tools for overhauling its carburetors. Some, such as the auxiliary-venturi removal tool, may be invaluable if you plan to overhaul many Weber carburetors. Other tools, such as reamers or gages, have special applications better left to an experienced technician. Improper use of these tools will quickly—and irretrievably—do more harm than good.

In practical terms, most repairs are made by individuals for their own vehicles, involving probably no more than three Weber carburetors at most (Ferrari owners humbly excepted). Moreover, home-bound "repairs" are most likely to be extensive cleaning and checking jobs—not overhauls in the usual sense of the word. Beyond a thorough cleaning, setting float level and replacing rubber parts and gaskets, comprehensive Weber overhauls should be better left to a specialist.

The low cost of a Weber carburetor argues against extensive or heroic repair procedures. Of course, some Webers are old and irreplaceable, but that fact is also an argument for repair by a factory-authorized facility.

If you're facing a Weber repair that you think requires a special Weber tool to save the carburetor, consult an authorized dealer.

phragm (not piston) accelerator pump, and such modern items as a power valve and fuel cut-off solenoid, not found on the IMPE. But the similarities between the IMPE and ICH are so overwhelming that the differences become negligible.

The DFT carburetor section details specific repair procedures for the emission controls that the ICH has, but the IMPE doesn't, so the ICH owner needs to read the DFT section, as well.

In looking at the chart of carburetor types, it's clear that the largest single group of Weber carburetors has two downdraft throats and two parallel throttle shafts. Two carburetors have been picked from this group: the DGAV as a representative of the generic type, and the DFT as a representative of all emission-controlled Weber carburetors. Both the DGAV and DFT have wide application.

A triple-throat IDA carburetor was chosen because it is a dramatic high-performance carburetor typically fitted to Porsche engines. There is both a two-throat and three-throat IDA carburetor: the two-throat type is similar to a side-draft DCOE stood upright. The three-throat IDA is its own design.

SINGLE-THROAT REPAIR: IMPE

The IMPE carburetor is a downdraft Weber with a single venturi. A manual choke and accelerator pump enrichens the fuel mixture as needed. The main casting includes the venturi and carries all the jets and their drilled passages, while the float bowl casting carries the float assembly.

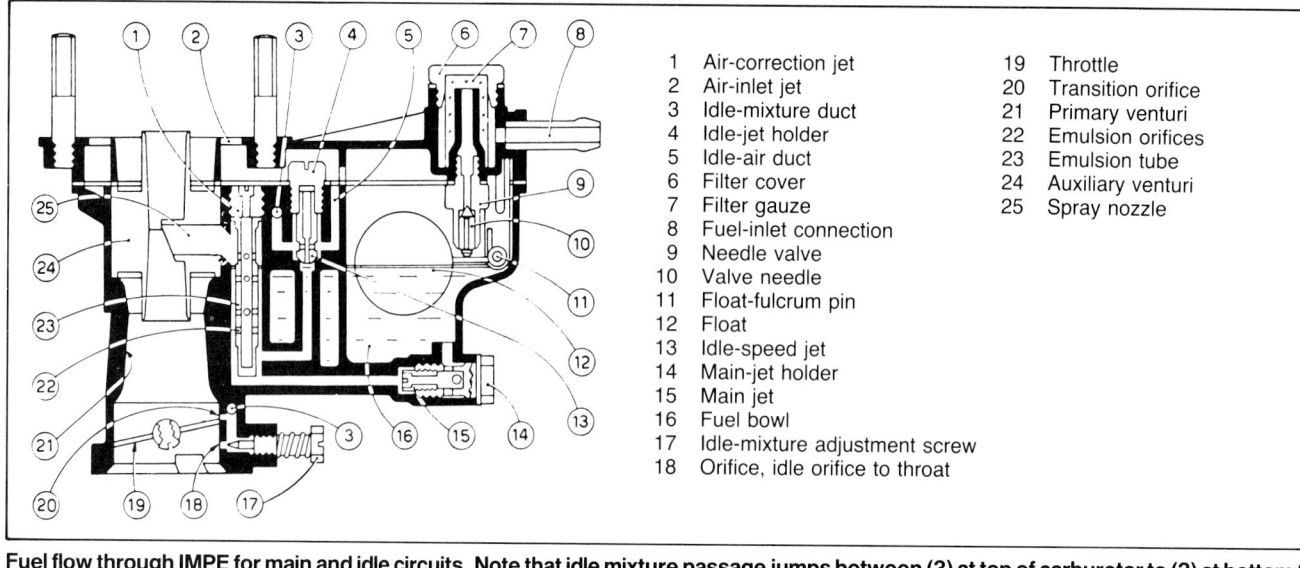

Fuel flow through IMPE for main and idle circuits. Note that idle mixture passage jumps between (3) at top of carburetor to (3) at bottom to suggest path of complete circuit.

This single-throat downdraft Weber is perhaps the least dramatic of all Weber carburetors, though it has been used on some interesting engines, most notably those of Carlo Abarth. The 32 IMPE was the stock carburetor for the 1100 Fiat. But it was most widely used on Abarth's derivations of the diminutive 600 Fiat, because that engine had a cast-in inlet manifold that dictated a single, downdraft carburetor.

The IMPE is fully tunable, with replaceable jets and venturis, so it is not a negligible carburetor even if it does represent the small end of Weber applications. Indeed, the IMPE was used on the 1-liter pushrod Monomille Abarth engine, an application that probably taxed its flow to the maximum.

FUEL FLOW

Fuel flows through the carburetor as shown in the above diagram.

DISASSEMBLY

It's not practical to disassemble this carburetor completely just to fix one part. You can do more accidental harm disassembling it than could ever result from normal wear on an engine. A little extra time spent diagnosing the problem can save the grief of damaging a perfectly good part during disassembly. The most common wear points on this carburetor are the throttle shaft bearing surfaces in the carburetor body. The IMPE uses the carburetor casting itself as the bearing surface for the throttle shaft. When these surfaces became worn, it was intended that the bearing bore be reamed oversize and an oversize shaft fitted.

Because the IMPE unit is an old application, its restoration is frequently tied in with the restoration of an entire vehicle. In such an instance, it is best to drill out the casting and insert bronze bushes to renew the throttle shaft bearing bore. This is a job for accomplished machinists only, because the throttle plates must be repositioned exactly over the progression holes, a setup that depends on the vertical location of the throttle shaft. So, any drilling or reaming operation must be done with the carburetor held in a drill vise and not moved. This will minimize the danger of drilling the bearing off-center.

The IMPE carburetor is unique in that its throttle valve has a hole drilled in it to pass idle air. Thus, at idle, the throttle plates are completely closed.

The procedure described below is divided into sections and covers the complete disassembly of the carburetor:

- Removal and disassembly of float bowl cover.
- Removal of components accessible when float bowl cover is removed.
- Disassembly of components on outside of carburetor.
- Removal of throttle shafts.

This is not the only order in which assemblies can be removed. Think of the carburetor as a collection of modules, any one of which can be worked on without disturbing the others. Here are the three main modules:

- Float bowl cover:
  This carries float, float-needle valve and fuel filter.
  Gives access to choke-circuit jet, air-correction jet and emulsion tube, accelerator pump and its delivery valve, choke valve and venturis.

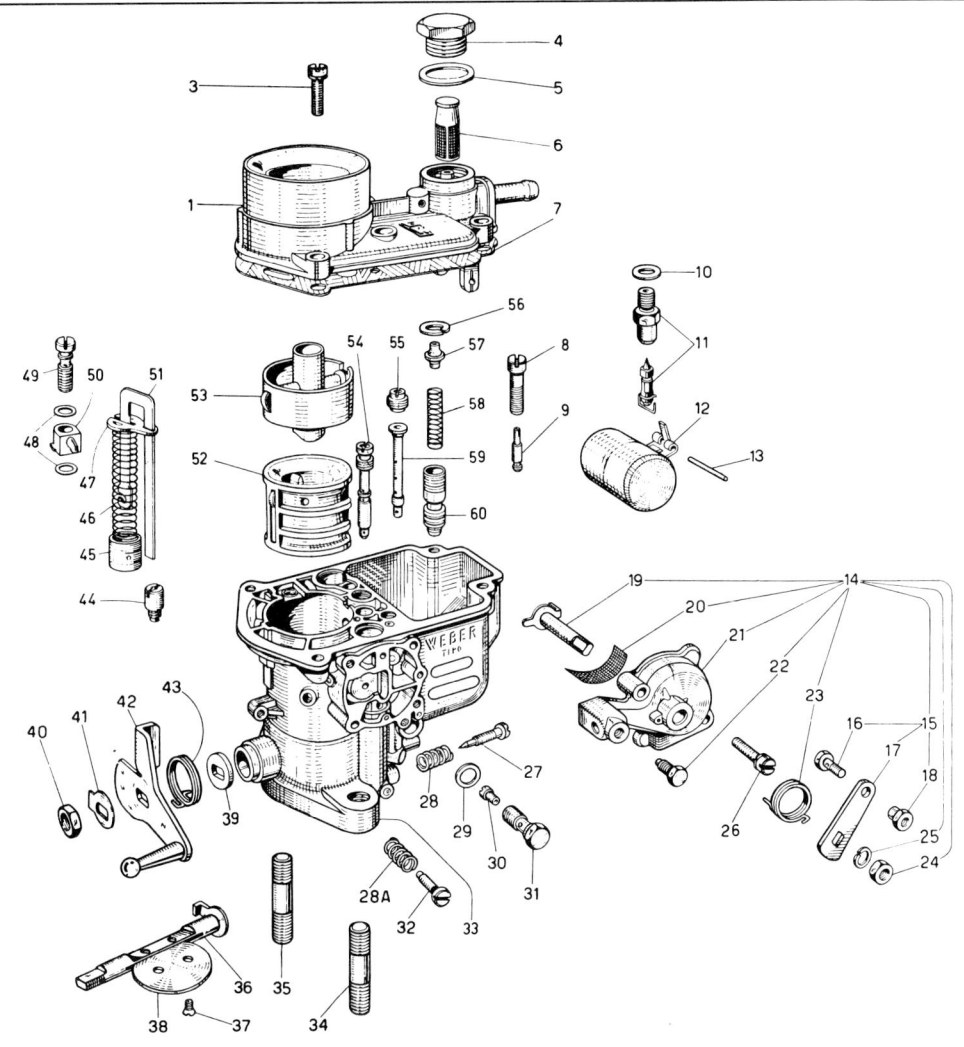

1	Float-bowl cover	20	Filter screen	37	Throttle-valve screw
3	Screw	21	Cover	38	Throttle valve
4	Filter-cap screw	22	Screw	39	Washer
5	Gasket	23	Choke spring	40	Throttle-shaft nut
6	Fuel filter	24	Nut	41	Throttle-shaft-end washer
7	Gasket	25	Washer	42	Throttle-shaft lever
8	Idle-jet holder	26	Choke-cover screw	43	Throttle-return spring
9	Idle jet	27	Idle-mixture screw	44-51	Accelerator-pump assembly
10	Gasket	28	Idle-mixture-screw spring	52	Venturi
11	Float needle valve	28A	Idle-speed-screw spring	53	Auxiliary venturi
12	Float	29	Washer	54	Idle jet
13	Float-fulcrum pin	30	Main jet	55	Air-correction jet
14	Choke	31	Main-jet holder	56	Retaining ring
15	Choke-lever assembly	32	Idle-speed screw	57	Spring guide
16	Screw	33	Carburetor body	58	Delivery-valve spring
17	Lever	34	Mounting stud	59	Emulsion tube
18	Nut	35	Mounting stud	60	Accelerator-pump delivery valve
19	Choke lever	36	Throttle shaft		

32 IMPE. Drawing courtesy Weber.

IMPE has fuel-filter screen under bronze cap screw. Blow screen clean with compressed air.

Float-bowl-cover removal: Don't remove protruding "screw" between float bowl and venturi.

Float-bowl cover lifts off with float assembly attached. Be careful not to bend float assembly.

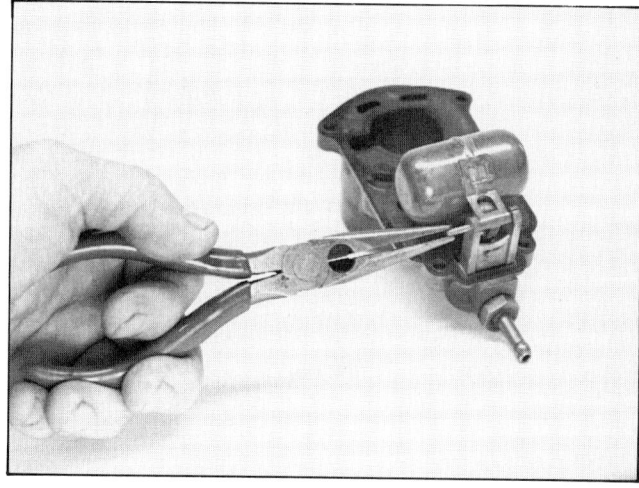
Float-pivot pin comes out easily with pliers. After float is removed, replace pin in cover to keep it safe. Always remove pin through solid boss, not split one.

- Exterior of carburetor:
  Idle-mixture adjusting screw
  Choke-actuating assembly
- Throttle shafts:
  Gives access to throttle plates.

Therefore, the following procedure recommends the order in which assemblies may be removed. A wide variation is possible, depending on the repair needs of the carburetor.

If you're going to do a complete overhaul, buy an overhaul kit. If you're doing anything less than a complete overhaul, try to save most of the old gaskets and diaphragms necessary to reassemble the carburetor successfully. To save a gasket or diaphragm, work very slowly and carefully to separate it from its mating surface.

Begin the overhaul by cleaning the outside of the carburetor as completely as possible. Scrape off baked-on dirt with a knife or screwdriver, but be careful not to gouge out pieces of the carburetor in an attempt to clean it. Use an aerosol can of carburetor cleaner to remove varnish and any remaining dirt. Follow the precautions and instructions on the can.

Float Bowl Cover—Remove the cap screw and washer that cover the fuel filter, then remove the filter and blow it clean with compressed air.

The float bowl cover is held on with

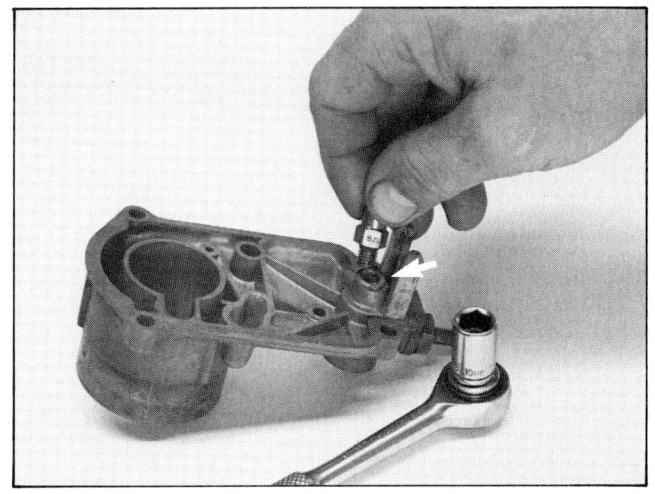
Use 10mm socket to loosen float needle valve. Don't ignore washer (arrow) that may be stuck to float-bowl cover.

Idle-jet assembly was "screw" that was sticking up through float-bowl cover.

Remove emulsion tube using small *E-Z Out* if stuck in bore. Work carefully to avoid marring tube.

Removing choke jet with correct screwdriver: Be careful not to damage jet during removal.

four screws. There is a fifth "screw" protruding through the cover close to the throat of the carburetor, but slightly inboard. That is the idle jet; don't remove it yet.

If you can't remove the cover with a slight pull, turn over the carburetor and use the handle of a screwdriver to tap the cover and loosen it from the carburetor body. Hit the cover lightly on the fuel filter extension. If you're trying to save the gasket for the float bowl cover, slide a knife along the gasket, separating it from the mating surface as slowly and carefully as possible. The gasket may be very brittle from age and heat; it's not always possible to remove one whole. If you tear the gasket, replace it.

Needle, Seat & Float—Turn the cover over, plug the filter opening with your finger, and blow into the fuel inlet. With only the weight of the float assembly pressing on the needle valve, you should not be able to blow air past the seat. If you can, and there's no dirt trapped between them, the needle and seat may have to be replaced.

Use needle-nose pliers to pull the pin that holds the float assembly to the cover. Remove the float and the float bowl cover gasket.

The needle valve is carried in its own holder; use a 10mm socket to remove it

Accelerator-pump piston is lifted out after its retaining cover is pried free.

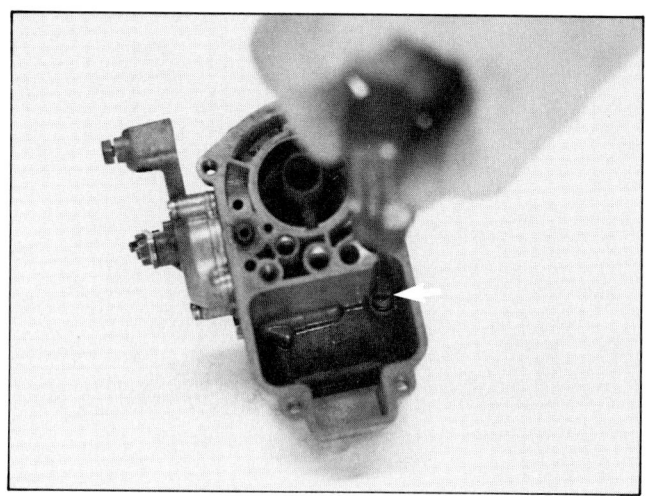

Accelerator-pump delivery valve sits at bottom of float bowl.

and its washer.

Jets—Remove the idle jet by unscrewing its holder and then pulling it free from the holder. Blow through the idle jet to verify that it is clear. Note: The jet is very small; hold it up to a light to verify that it is clean.

Remove the air-correction jet and withdraw the emulsion tube beneath it.

Remove the choke jet, being careful to use a screwdriver that just fits the slot in the jet. Grind a screwdriver to fit if you do not have one that will remove the jet without damaging it.

Accelerator Pump—The accelerator-pump piston is secured to the carburetor body by a U-shaped cover with a snap fit. Carefully pry up the cover and withdraw the entire assembly. The piston itself can be removed, if necessary, by driving out the pin that holds the piston to the flat shaft.

Remove the accelerator-pump delivery valve from the bottom of the float bowl. Blow through both sides of it to verify one-way operation.

Undo the accelerator-pump delivery valve attaching screw and remove it with the delivery nozzle. There is a washer above and below the nozzle.

With a small screwdriver, pry up the choke valve cover. Remove the cover, spring holder below it and the spring.

Venturi—Pull the auxiliary venturi free of the carburetor body. If it is stuck, you may have to remove the throttle shaft in order to fit a correctly sized wood dowel from beneath to drive it out. Remove the venturi next.

Choke Assembly—On the outside of the carburetor, remove the two screws attaching the choke assembly cover to the carburetor body.

Remove the screen from the choke assembly cover and clean it thoroughly. The choke mechanism usually doesn't need to be disassembled. As a general rule, avoid disassembling the choke mechanism; it can be tricky and time consuming to put back together. A thorough cleaning is usually all that's needed.

If necessary, disassemble the choke lever assembly by removing, in order, locknut, washer, arm and spring.

Then, pull the choke shaft from the body of the cover. Remove the choke valve from inside the carburetor body.

Main Jet—Remove the main jet holder with its washer. Unscrew the main jet from the holder.

Idle Mixture Screw—Count the number of turns required to seat the idle mixture richness screw, then back it out all the way, removing it and its spring. Unscrew the idle *adjustment* screw and remove it with its spring.

Throttle Valve & Shaft—Don't remove the throttle valve and shaft unless **absolutely necessary.** What's absolutely necessary? If the valve is binding and can't be freed with cleaner, the shaft is bent or worn. A worn throttle shaft will produce an erratic idle speed or an inconsistent return to a stable idle. If faced with these conditions, then consider replacing the complete unit. The removal and assembly procedures require a stout heart, ample dexterity, and plenty of time.

First, remove the two screws holding the throttle valve to the shaft. The ends of the screws are punched or *staked* to keep them from vibrating loose. Apply only *slight pressure* to the screws with a screwdriver to avoid bending the throttle shaft. Work the screws out carefully in 1/2-turn increments alternating with 1/4-turns back. If this technique doesn't remove the screws, grind off their ends so they can be removed without stripping the threads in the throttle shaft. The shaft itself is relatively hard, and even if you

Accelerator-pump nozzle comes off after screw attaching delivery valve is removed.

Auxiliary venturi slips out of bore. If stuck, work very carefully to remove it. Cast metal is easily damaged.

do strip a screw, there will be enough thread left in the shaft to chase with a 4 X 0.7mm tap so new screws can be used.

Mark the throttle valve so it can be reinstalled exactly as it was removed, and then remove the valve.

Bend back both ends of the lock tab and use a 12mm wrench to loosen the nut on the end of the throttle shaft.

The shape of the parts that you'll remove from the throttle shaft will change depending on the application. Generally, the next step is to strip the shaft of all its components. Lay them out carefully so you can reassemble them exactly in reverse order of how they came off. A simple drawing of their order could be invaluable at reassembly.

Begin by removing the nut, then slip off the lock-tab washer, main arm, spring and two seals (if used). There is a seal on both ends of the throttle shaft. The one near the accelerator-pump arm may appear to be part of the throttle shaft. Pull the throttle shaft free from the carburetor body from the side opposite the threaded end.

INSPECTION
Cleaning—The first step in inspecting a carburetor for wear is to clean all its parts thoroughly. Use an aerosol spray can of carburetor cleaner, or soak them in a large can of solvent designed specifically for carburetors. Follow the directions and precautions on carburetor cleaners. Because they are so strong, use goggles and rubber gloves when handling them.

If soaking carburetor parts, be sure they won't be damaged by the strong cleaner. Some plastic and rubber parts won't survive a long dunking.

Use a brush to swab cleaner on large surfaces to dissolve deposits. And an ear syringe makes a good tool to shoot cleaner through passages. *Never use a wire or drill to clean passages.* The slightest mark can change passage's flow characteristics and upset metering.

Inspect all of the drilled passages in a carburetor as you clean them with the aerosol spray or ear syringe. Wear goggles when you do this. Work in a well-ventilated area away from fire or pilot lights—such as on a gas range, water heater or furnace.

Aerosol cans come with a plastic tube that inserts into the spray nozzle. Use it to probe the passages while spraying cleaner. It's unusual for passages to be blocked. Dirt usually blocks the jets before the passages. A more common problem to look for is the loss of plugs used to block the ends of drilled passages.

One of the marvelous features of Weber carburetors is that they have well-defined fuel passages. Spend enough time to identify all the drilled passages; figure out where they begin and end, and how fuel flows through them to get where its supposed to.

Float Bowl Cover—The gasket sealing the float bowl cover to the carburetor main casting should show that the sealing surfaces of the two castings are parallel. Check the float bowl cover for broken flanges, cracks or other physical damage. Check the threads for the fuel filter plug and threads for the power valve.

Check the fuel filter for any damage to the screen. Blow it clean with compressed air to remove any dirt.

The floats should not show any evidence of leaking. Shake the assembly near your ear and listen for any gasoline sloshing inside. If the floats leaked, replace them.

The float needle valve and its seat should not show uneven wear where they seat against each other.

Jets—Blow compressed air through the idle jet, air-correction jet, choke jet, emulsion tube and accelerator-pump jet

Venturi, held in place by auxiliary venturi, is easily removed.

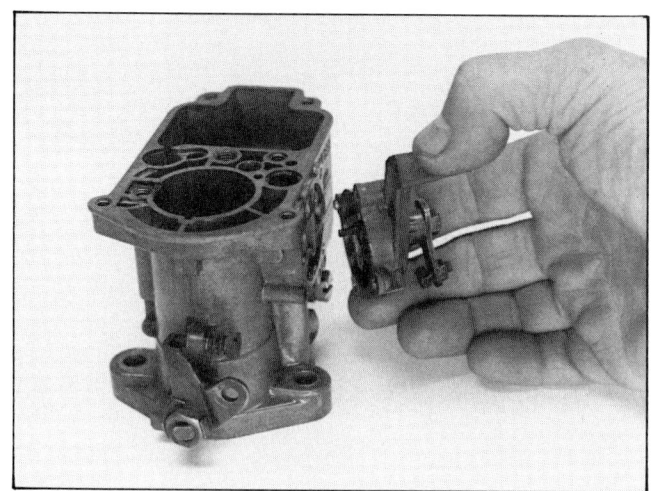
Lift off choke assembly. Don't take it apart unless it is broken.

Remove choke valve. Examine it for scoring. Valve should move easily in its bore.

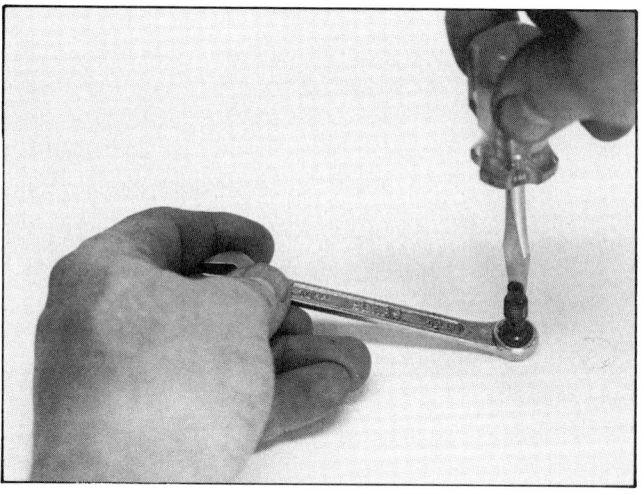
Secure main-jet holder with 10mm wrench, then carefully unscrew main jet from holder.

to clean them. Never use a wire to probe or clean the jets. Blow through the accelerator-pump delivery jet from both sides; it should be a one-way valve.

If the auxiliary venturi has been removed, blow though the passage and remove any burrs or dirt on its surface.

Main Casting—Check for any damage to the main casting. Use a straightedge to check the mating surface of the carburetor's base for warpage.

Check the main jet for blockage by holding it up to a light. Blow through the jet with compressed air if it's clogged. The choke piston should not show scoring or wear, and should fit in its bore smoothly.

Throttle Shaft—The throttle shaft should be straight, with no visible wear where it seats in the carburetor body. Roll it on a flat surface to check its straightness. Double-check the threads in the end of the primary throttle shaft and the threads on both shafts where the throttle valves are attached.

Fit the bare shaft (without seals) to the carburetor body and check for radial play. If you detect play, the bearing bore must be renewed. Note: The symptom for a worn throttle shaft or bore is an erratic idle speed or inconsistent failure to return to correct idle speed. Additionally, the throttle butterfly valves

Remove idle mixture screw. Count turns needed to seat screw, then back it out. Note number for initial setting at reassembly.

Dangerous work: removing throttle plate screws requires care, some luck. Don't remove screws or throttle plate unless absolutely necessary.

Throttle plate slips out easily. Note *exactly* the position of its beveled edges, and put it back the same way.

Disassembled throttle shaft with parts laid out in order of assembly along shaft.

should have smooth edges and be flat. The throttle shaft spring should be strong and not deformed.

REASSEMBLY

The disassembly of the IMPE is, in itself, a lesson in reassembly. After all the components have been inspected and damaged parts replaced, begin reassembling the major components first. The disassembly photos will be helpful during assembly. Start with either the float bowl cover or main casting. The notes below give the most important points to remember for reassembly.

Throttle Shaft—Install one seal onto the accelerator-pump-lever end of the throttle shaft, then fit the other seal into the body of the carburetor. Make sure it seats correctly. In both cases, the solid face of the seal must install *toward* the throttle valve. Install the throttle shaft and then assemble the following parts to it: spring, main arm, lock-tab washer and nut, in that order.

Installing the throttle shaft spring is a challenge in manual dexterity. Fit its straight end into the carburetor body to locate it. Then, slip the main arm into the hooked end of the spring and wind the spring about 3/4 turn. At this point, you should be able to slip the arm onto the throttle shaft. Be careful not to trap the

Assembled shaft is tidy, considering number of components involved. Getting return spring on may require a helper. Bend lock tab (arrow) on washer over nut flat to secure nut.

Throttle plate in place: Work throttle shaft back and forth a bit to make sure plate is seated against throttle bore. Then secure throttle plate screws by *staking* them.

spring between the arm and carburetor body. Small needle-nose pliers or a screwdriver will help keep the spring coils free as you fit the arm. When the spring is in place, fit the lock-tab washer and nut, then recheck that no spring coil is trapped behind the arm. Tighten the nut and bend the lock tabs over it.

Throttle Valve—There are two dangers in reassembling the throttle valve: bending the throttle shaft and not positioning the valve correctly. The first danger requires only care. Because the IMPE has a soft-metal throttle shaft and relatively soft screws, use much less torque during reassembly of this carburetor than for those with hardened-steel shafts and screws.

Seating the throttle valve correctly requires some strategy. The holes in the throttle valve are oversize to permit the valve to be positioned for a perfect seat in the bores. The auxiliary venturi should be removed to give free access to the attaching screws during this operation.

Put the throttle valve in place exactly as it was removed. There is usually a number stamped on the valve. The number *faces down* toward the carburetor base. Let the valve position itself in the bore, apply fuel-resistant Loctite to the screw threads, and tighten the attaching screws lightly, being careful not to move either the shaft or valve.

Double-check your work by holding the carburetor up to a bright light and looking around the circumference of the valve. There should be a light-tight seal between the valve and bore.

Tighten the screws securely and then re-check your work. The valve is held in place by the tightness of the screws, but the ends of the screws should be deformed (staked) to ensure they don't vibrate loose and drop into the engine.

Take the steps outlined to prevent this catastrophe. Use fuel-resistant Loctite on the threads and deform the screw ends. You can deform each screw end with a punch only if you firmly support the opposite end as follows so no force is transmitted to the throttle shaft.

Set a 1/4-in. drive, 1/2-in. socket on a steel or concrete surface with the square drive end facing up. Place the carburetor on top of the socket so a throttle valve screw head rests against the smooth surface of the socket. Then, place a punch down the venturi and position it on the end of the screw that's being supported by the socket. The valves won't allow the throttle shaft to rotate so you can hit the screws dead-center, but the angle is so small that it won't matter. Try to deform the screw end with a single sharp blow as closely centered and aligned to the screw as possible.

You can also secure the screws with Vise-Grip pliers. Reach inside the venturi with a small pair of them to squish the threads on the exposed screw end. This will deform the end and keep the screw from backing out.

Carburetor Body—Reassemble the idle adjustment screw and its spring. Screw it into the carburetor so the end of the screw just touches the throttle arm.

Reinstall the main jet into its holder and snugly fit the holder and its washer to the carburetor body.

Refit the idle mixture adjustment screw, seating it lightly, then back it out the number of turns noted during disassembly. As a guide, 1-1/2 turns out will work if the baseline has been lost.

Choke Assembly—Slip the choke valve into the carburetor (pointed-end down), then install its spring and spring holder. Snap the choke valve cover in place. Double-check that the cover is flush with the surface of the carburetor body.

Place the screen in the choke assembly cover. Be sure the edge of the screen is

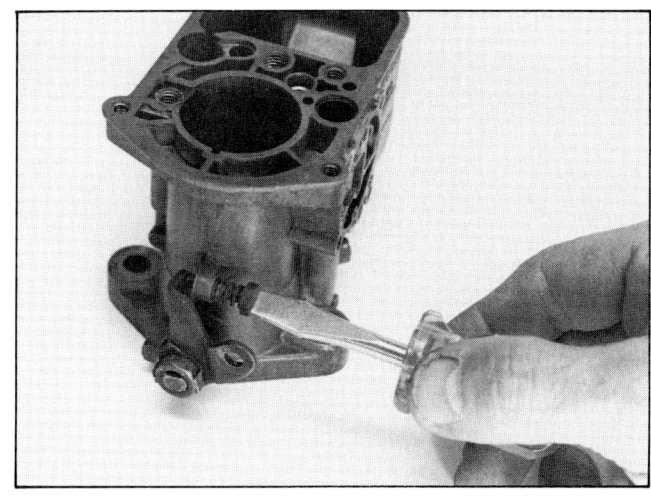

Reassembling idle speed screw. Screw should just touch throttle lever.

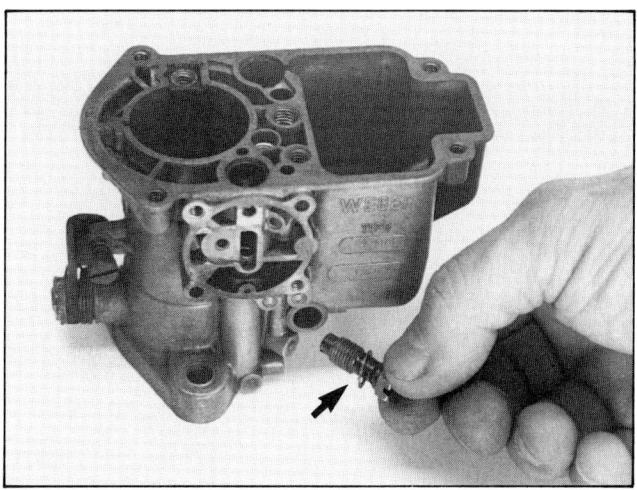

Main-jet installation: Don't forget copper washer; it assures an air-tight seal.

flush with the mounting surface of the assembly, then reassemble the elements of the cover if you disassembled them. The order is: spring, arm, washer and locknut.

Then, engage the choke valve in the carburetor body with the tab on the arm of the choke assembly. Attach the assembly to the carburetor body with the two shoulder screws, then operate the choke to make sure nothing binds.

Venturi—Slip the venturi and auxiliary venturi in place. The number on the venturi goes up, and the longest part of the auxiliary venturi points up.

Accelerator Pump—Refit the two washers to the accelerator-pump delivery nozzle and screw it into place. Refit the accelerator-pump delivery valve in the bottom of the float bowl.

Reassemble the accelerator-pump piston and spring onto the arm if you disassembled it. Fit the piston into its bore on the carburetor, then snap the U-shaped retaining cover in place.

Jets—The choke jet is the long one-piece tube with a large-diameter hole and screwdriver slot at the top. Screw it into the carburetor body. Drop the emulsion tube into its bore and fit the air-correction jet over it. Reassemble the idle jet to its holder and screw the holder into the carburetor body. See the disassembly photos for help here.

Needle, Seat & Float—Refit the float needle valve into its holder with the small steel ball visible. Slip a washer onto the holder, then screw the holder into the float bowl cover. Check to see that the needle valve moves freely. Refit the float bowl cover gasket. Place the float in place and pin it. Reattach the float bowl cover with four screws, then replace the filter and cover it with its cap screw and washer.

IMPE INITIAL SETUP & TUNING
Cold Engine:
- If float level has not been set, do it now.
- Check for fuel leaks. If using an electric fuel pump, switch on ignition so fuel pump operates. Then check all fuel fittings for leaks. If mechanical pump is used, disconnect high-tension (large) wire from coil, crank engine for several seconds, then check for fuel leaks. Reconnect coil wire.
- Remove air filter, locate accelerator-pump outlet in venturi. Open throttle, then close it; fuel should flow from accelerator-pump outlet. If it doesn't, correct problem.
- Have a helper fully depress accelerator pedal. Verify that throttle plate opens fully. If not, readjust linkage so it does.
- Verify, if necessary, that idle speed and idle-mixture-richness adjustment screws are at their initial settings. Connect tachometer to engine to monitor idle rpm. Start engine and let it warm to operating temperature. This may require resetting idle speed so engine continues to run.

Warm Engine:
- Adjust idle speed to approximately 850 rpm. Less than 1000 rpm is OK if intermediate or main circuits are not operating and engine continues to run.
- Turn idle-mixture adjustment screw until engine idles at maximum speed. Experiment to obtain this setting. Start by turning screw out in equal 1/2-turn increments—rotate slot in screw's head exactly 180°—until engine rpm begins to decrease. If turning screw out only increases engine rpm, reset screw to its initial setting and readjust the baseline idle speed so it is as low as possible. Then, repeat the idle-mixture adjustment. When engine rpm begins to decrease, turn in idle-mixture screw equally until maximum idle speed is obtained.
- Readjust idle speed to 850 rpm, or as specified in owner's manual or underhood label.
- Disconnect tachometer.

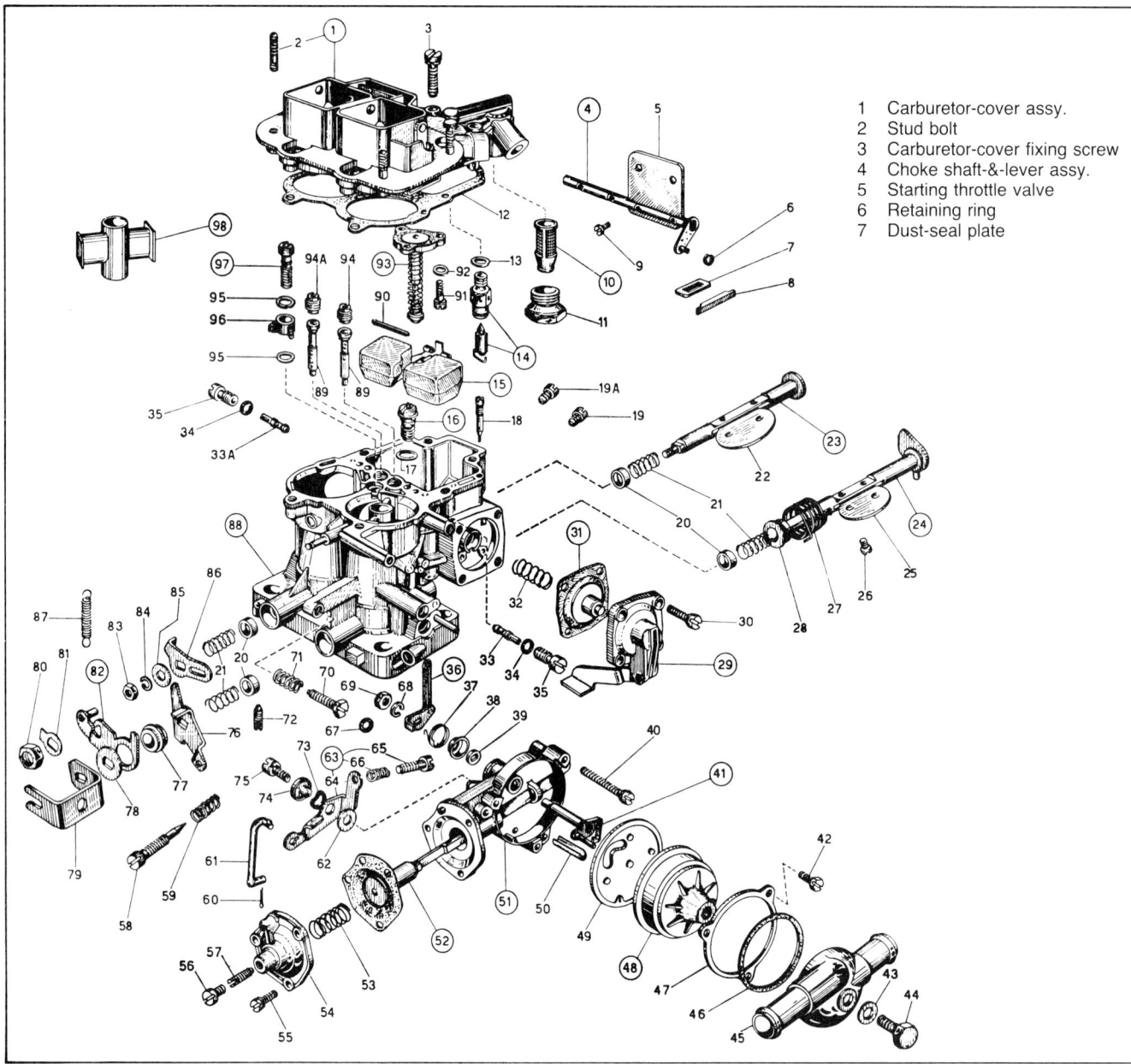

1 Carburetor-cover assy.
2 Stud bolt
3 Carburetor-cover fixing screw
4 Choke shaft-&-lever assy.
5 Starting throttle valve
6 Retaining ring
7 Dust-seal plate

32/36 DGAV. Drawing courtesy Redline, Inc.

DUAL-DOWNDRAFT REPAIR: DGAV

The DGAV carburetor is representative of the family of downdraft Weber carburetors with two venturis and two parallel throttle shafts. The dual-venturi, vertical carburetor is standard equipment on most Italian cars that do not use horizontal Webers such as the DCOE. It is a favorite conversion carburetor for every vehicle from Austin to Volkswagen.

The DGAV uses a progressive linkage from the primary throttle shaft to operate the secondary throttle, and there is an automatic choke, power enrichment valve and accelerator pump to enrich the fuel mixture as needed. The main casting includes the two venturis and carries all the jets and their drilled passages. The float bowl cover casting carries the power enrichment valve and float assembly.

DISASSEMBLY

See the sidebar, page 45, for information on special Weber tools. It's not practical to disassemble this carburetor completely just to fix one part, because you can do more accidental damage than could ever result from normal wear. A little extra time spent diagnosing the problem can save the grief of breaking a good part during disassembly.

8	Dust-seal plug
9	Choke-plate fixing screw
10	Strainer assy.
11	Strainer inspection plug
12	Carburetor-cover gasket
13	Needle-valve gasket
14	Needle-valve assy.
15	Float assy.
16	Full-power needle-valve assy.
17	Power-valve gasket
18	Pump-discharge blanking needle
19	Primary main jet
19A	Secondary main jet
20	Shaft retaining bushing
21	Bushing retaining spring
22	Secondary throttle valve
23	Secondary-shaft assy.
23	Secondary-shaft assy., oversize
24	Primary-shaft assy.
24	Primary-shaft assy., oversize
25	Primary throttle valve
26	Throttle-plate fixing screw
27	Shaft return spring
28	Spacer
29	Accelerator-pump-cover assy.
30	Pump-cover fixing screw
31	Accelerator-pump-diaphragm assy.
32	Pump loading spring
33	Primary idle jet
33A	Secondary idle jet
34	Gasket, idle-jet holder
35	Idle-jet holder
36	Choke-control-lever assy.
37	Spring, fast-idle cam
38	Spring, retaining cover
39	Washer, shaft
40	Choke fixing screw
41	Auto-choke shaft-&-lever assy.
42	Plate screw
43	Washer, water-cover fixing screw
44	Water-cover fixing screw
45	Auto-choke water chamber
46	Water-chamber-seal gasket
47	Thermostat-assy. locking ring
48	Auto-choke thermostat assy.
49	Gasket, auto-choke body
50	Plate, choke shaft
51	Auto-choke-body assy.
52	Choke-diaphragm assy.
53	Diaphragm loading spring
54	Auto-choke cover
55	Auto-choke-cover fixing screw
56	Screw plug
57	Diaphragm adjusting screw
58	Idle adjusting screw
59	Spring, idle adjusting screw
60	Split pin
61	Fast-idle control rod
62	Washer, loose lever
63	Fast-idle loose lever assy.
64	Lever
65	Screw
66	Spring
67	Auto-choke O-ring seal
68	Spring washer
69	Throttle-shaft fixing nut
70	Primary-throttle adjusting screw
71	Spring, throttle adjusting screw
72	Secondary-throttle adjusting screw
73	Wave-washer, loose lever
74	Bushing, loose lever
75	Loose-lever fixing screw
76	Primary-throttle control lever
77	Bushing, loose lever
78	Washer, loose lever
79	Throttle-valve control lever
80	Throttle-shaft fixing nut
81	Lock washer
82	Loose-lever assy.
83	Secondary-shaft fixing nut
84	Spring washer
85	Washer, loose lever
86	Secondary-throttle control lever
87	Spring, loose lever
88	Carburetor body
89	Emulsion tube
90	Float fixing pin
91	Control-valve retaining screw
92	Washer, control-valve screw
93	Power-valve assy.
94	Primary air-correction jet
94A	Secondary air-correction jet
95	Pump-jet gasket
96	Accelerator-pump jet
97	Pump delivery-valve assy.
98	Auxiliary venturi

The most common wear points on this carburetor are the throttle shaft bores in the carburetor body. The DGAV has nylon seals around the shafts where they pass through the main casting. Replacing these seals means removing the throttle plates, which is the most challenging repair operation for this unit. If the shafts are not worn, refurbishing a DGAV is straightforward enough that it can be performed without removing the carburetor from the car.

The procedure described below is divided into sections and covers the complete disassembly of the carburetor when it's off the car:

● Removal and disassembly of float bowl cover.

● Removal of components accessible when float bowl cover is removed.

● Disassembly of components on the outside of carburetor.

● Removal of throttle shafts.

This is not the only order in which assemblies can be removed. Think of the carburetor as a collection of modules, any one of which can be worked on without disturbing the others. Here are the three main modules:

● Float bowl cover:
Carries choke shaft and butterfly valves, power valve assembly, floats, float needle valve, fuel filter. Gives access to main jets, air-correction jets and emulsion tubes, accelerator-pump delivery valve, full-power needle valve, and auxiliary venturis.

● Exterior of carburetor:
Idle jets
Accelerator pump
Automatic choke assembly

● Throttle shafts:
Gives access to shaft seals and throttle plates.

The following procedure suggests the order in which assemblies may be removed. Variations are possible, depending on the repair needs of the carburetor.

If you're going to do a complete overhaul, buy an overhaul kit. If you're doing anything less than a complete overhaul, you may be able to save most of the old gaskets and diaphragms necessary to successfully reassemble the carburetor. If you're trying to save a gasket or diaphragm, work very slowly and carefully to separate it from its mating surfaces.

Begin the overhaul by cleaning the outside of the carburetor as completely as possible. Scrape baked-on dirt with a knife or screwdriver, but be careful not to gouge out pieces of the carburetor in an attempt to clean its surfaces. Use an aerosol can of carburetor cleaner to remove varnish and any remaining dirt. Follow the precautions and instructions on the cleaner's can.

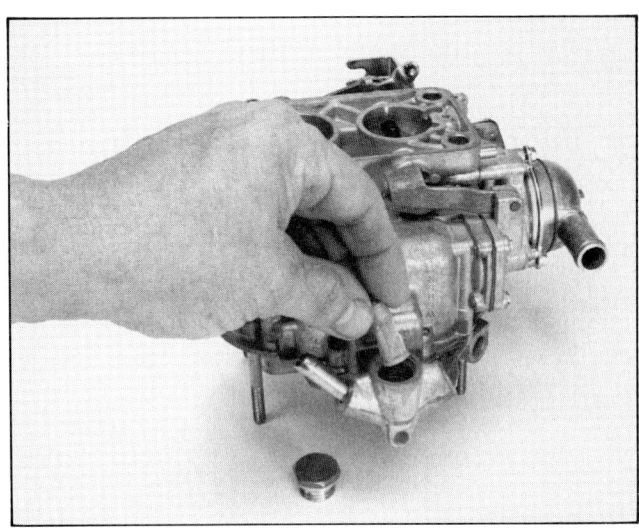
DGAV fuel filter is beneath fuel-inlet boss on float-bowl cover. To clean, blow from inside out with compressed air.

E-clip holds choke control arm to choke assembly in float bowl. Remove by pressing off with screwdriver. Be careful not to lose clip by pressing too hard, too fast.

Float-bowl cover removal: Note that screwdriver blade is ground to fit screws perfectly. It's important to keep screwdriver from slipping, marring screws.

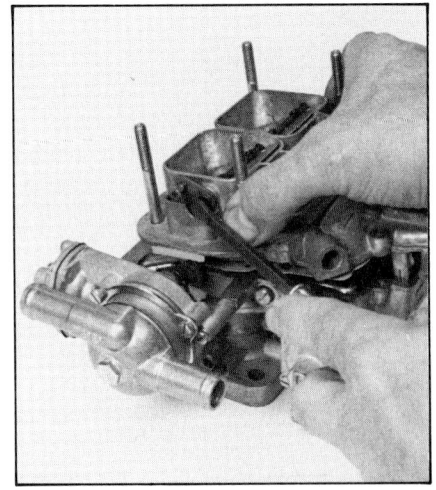
Float-bowl screws removed, and cover lifted free: Use screwdriver to release choke control arm.

Float Bowl Cover—Turn the carburetor over and use a 14mm wrench to loosen the fuel filter plug. Lift out the fuel filter, clean it with carburetor cleaner and set it aside.

The only link between the main body of the carburetor and float bowl cover is the arm that controls the choke plate. It's held onto the choke plate bellcrank with an E-clip. *Very carefully* force back the clip with a screwdriver blade and remove it. This piece is easy to lose, so store it appropriately. Don't try to disconnect the arm from the bellcrank yet.

Unscrew the six screws attaching the float bowl cover and loosen the cover. If you can't remove the cover with a slight pull, turn over the carburetor and use the handle of a screwdriver to tap the cover and loosen it from the carburetor body. Hit the cover lightly on the fuel filter extension. If you're trying to save the gasket for the float bowl cover, slide a knife along the gasket to separate it from the mating surface. Do this slowly and as carefully as possible. Don't nick or gouge the mating surfaces. The gasket may be brittle from age and heat. It's not always possible to remove one whole, so if you do tear the gasket, replace it.

Lift the float bowl cover free from the carburetor body, and slip the arm off the choke-plate bellcrank. Turn over the cover.

Needle & Seat—Plug the fuel filter opening with your finger, and blow into the fuel inlet. With only the weight of the float assembly pressing on the needle valve, you should not be able to blow air past the valve. If you can, and there's no dirt trapped between them, the needle and seat should be replaced.

Use needle-nose pliers to pull the pin holding the float assembly to the cover. Don't pull the pin through the *split* boss; pull it through the other side. The float assembly will lift off, taking the needle valve with it. The needle valve is attached to the float assembly with a loop of spring wire. Don't bend the wire as you slide it from the float.

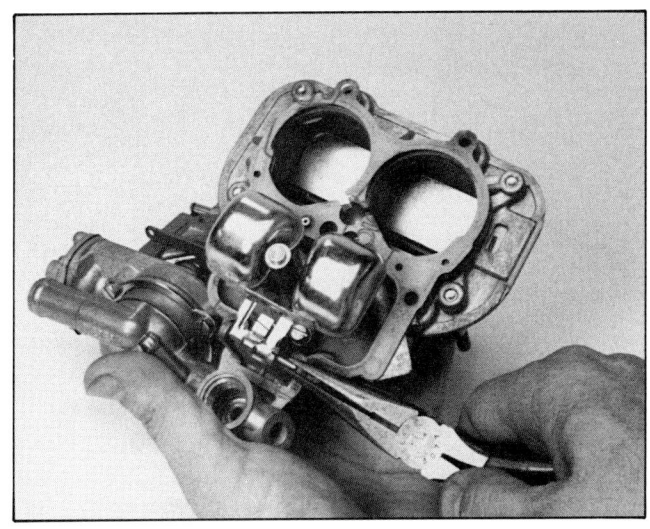
Remove float-pivot pin. Needle valve is attached to float assembly; be careful not to drop it when removing float.

Remove float-bowl-cover gasket. Gasket can be reused if not torn or leaking. Don't use liquid gasket material. It's too messy for this joint and will clog power-valve passage.

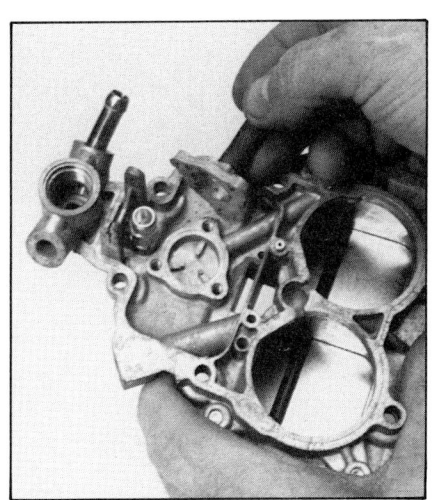
Power valve is held to float-bowl cover with three screws. Remove valve carefully so diaphragm doesn't tear. Power valve can usually be left intact.

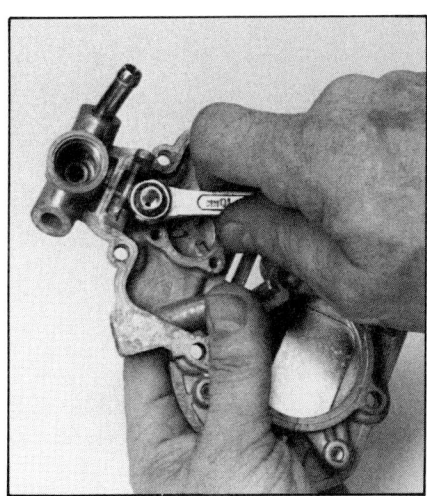
Needle-valve seat is removed with 10mm wrench. Be sure to save metal gasket under seat.

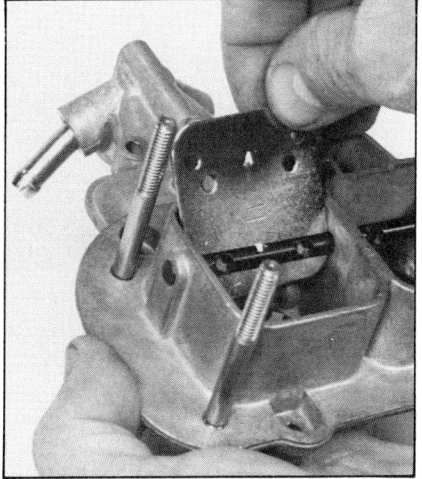
Removing choke valves (plates) easily damages shaft. Avoid disassembly unless absolutely necessary. Note identifying white paint mark applied to shaft and valve to ease matching parts at reassembly.

Power Valve—The gasket usually stays attached to the float bowl cover. Carefully lift it off. It's preferable to leave the power valve in place unless its diaphragm is damaged. If it's necessary to inspect or replace the diaphragm, unscrew the three screws for the power valve assembly and separate the power valve from the cover. Be careful not to tear the diaphragm. Check to make sure the diaphragm is flexible and doesn't have any cracks or tears.

Needle-Valve Seat—Unscrew the needle-valve seat using a 10mm wrench. This step is necessary only if you're replacing the needle and seat as a set. The gasket under the valve seat is shiny, and can look like it's part of the casting. Double-check to be sure you've removed the gasket. If you don't and install another one on top of it during reassembly, the float level will be lowered significantly, and lean the mixture.

Choke Plates—Unless **absolutely necessary**, don't remove the choke plates. It's easy to damage the threads in their shaft. A thorough cleaning and lubrication is usually all that's required to renew them. If they were binding or their shaft is bent, remove them.

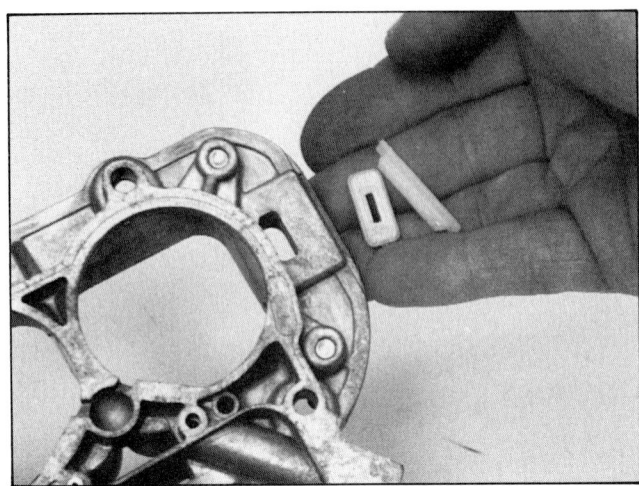
Choke-arm dust seals are held in casting on underside of float bowl. Pry off longer seal: O seal will fall out. Both seals can be re-used.

Remove main jet. Screwdriver must fit slot correctly or jet could be damaged. Jets fit at bottom of float bowl. Entire main circuit is illustrated here. Air-correction jets sit on top of cast-in wells, with main jets at bottom. Note how auxiliary venturi webs point at wells. Drilled passage in webs directs fuel from top of wells to venturis.

Remove power-valve plunger assembly from bottom of float bowl. Note that screwdriver is large, fits slot.

Air-correction jet removal: Use screwdriver that fits; don't damage screw slot.

Mark the choke valves so you can replace them exactly as they were removed. Then, try to unscrew the four screws holding the two valves to the shaft. The screw ends are punched (staked) to keep them from vibrating loose. If you can't unscrew them with just a little torque, grind away the damaged threads with a small grinding wheel. Once the screws are out, slide the plates from the shaft.

Pull the choke shaft from the casting and then remove the two dust seals.

Jets, Power Valve & Emulsion Tubes—Two main jets are in the bottom of the float bowl. Carefully remove them with a screwdriver that snugly fits their slots. If the screwdriver isn't a good fit and slips, the flow pattern of the jet could be changed as a result of any damage.

Unscrew the power-valve plunger assembly from the bottom of the float bowl and remove it with its gasket.

The two air-correction jets are on the top of the main casting. Unscrew them to get at the emulsion tubes trapped below the jets.

The emulsion tubes will probably be stuck in their bores. Remove them with a 3mm tap. Cut three or four threads at the top of the tube with the tap and then pull on the tap.

Remove the idle jet holders. Check the

Use wire to remove emulsion tube. Pry against tube and lift. If stuck, make about three threads in tube with 3mm tap and pull tube free with tap.

Idle-jet removal: Another idle jet is on opposite side of carburetor. Rubber O-ring seals jet assembly.

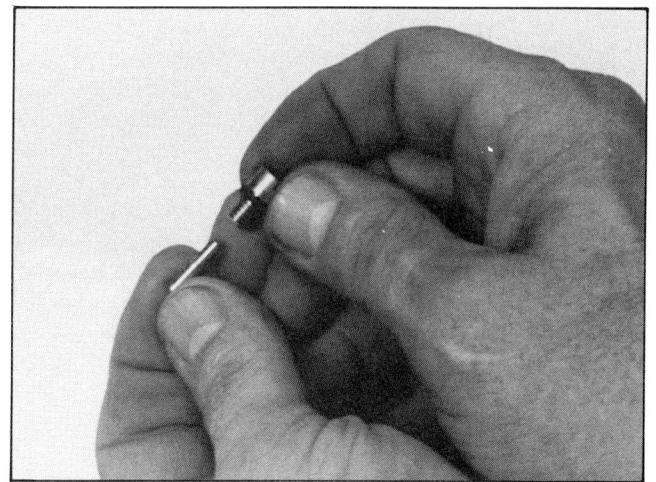
Idle jet pulls apart for cleaning, if necessary. Never use pliers to separate jet from holder.

Remove idle mixture screw. First, record number of turns to seat, then unscrew. Use same number for initial setting at reassembly.

rubber gasket sealing the holder to the body of the carburetor. It should be pliable and not broken. The idle jets can be separated from their holders simply by pulling them from the threaded holders. Blow through the jets to make sure they're clean.

Idle Mixture Screw—Count the number of turns required to seat the idle mixture screw, and unscrew it from the carburetor body. When you replace the screw, use the number of turns from the seated position to make the initial idle mixture adjustment. Remove the idle speed adjustment screw.

Accelerator Pump—The accelerator pump cover is held to the body of the DGAV with four screws. These screws have tapered seats and no lock washer, and are usually a little hard to remove. Be sure the screwdriver blade fits tightly in the screw slot, then carefully remove the screws. There is a rubber diaphragm sealed by the cover. When you pull the cover back, be very careful not to tear the diaphragm.

A gasket will be stuck to the diaphragm on one side. Leave it alone unless it's seriously damaged. Check the part of the diaphragm which moves to see that it's not cracked. As you remove the diaphragm, you'll find a small spring underneath it. Set it aside.

Remove idle-speed screw. Removing it is usually unnecessary, unless total overhaul of carburetor is required.

Remove accelerator-pump cover. Screws are typically hard to remove, and diaphragm beneath cover is easily torn. Don't disassemble unless pump is defective.

Accelerator-pump diaphragm and spring are held in body of carburetor. Diaphragm rubber should be supple, tear-free. At reassembly, replace diaphragm exactly as removed.

Pump jet also holds accelerator-pump-circuit nozzles. See that check ball operates freely.

The next step in removing the accelerator pump is to remove the jet that sits between the two venturis. It's retained by the *pump-delivery valve* assembly, which looks like a large screw from the top, but contains the pump circuit's one-way valve. After you unscrew the delivery valve assembly, you'll find two thin washers sealing the jet and delivery valve to the carburetor body. Make sure the hole in the jet is clear, and keep the entire assembly together by reassembling it on the bench.

The final step in dismantling the accelerator pump circuit is to remove the *blanking screw* from the carburetor body. Store the screw with the jet and delivery valve.

Venturis—Remove the auxiliary venturis by lifting up on them. If it's necessary to remove the auxiliary venturis and they won't lift out, tap them out using a wood dowel from underneath. Weber has an expensive tool for doing this if you can't remove them with a wood dowel.

Automatic Choke—The operation of the automatic choke is adjusted by rotating the water chamber assembly in relation to the choke body. The three screws you're about to loosen permit changing this adjustment. The factory punches an *index mark* to indicate the relationship between the two castings. The mark is usually on the top of the choke assembly and overlaps the joint where the water chamber mates with the choke body. Find the index mark and be prepared to reassemble the two halves in exactly the same relationship. Make your own reference mark if there's any question about the relationship of the two parts.

Remove the three bolts holding the automatic choke water chamber assembly to the choke body. Use a 7mm wrench to remove the bolt covered by the water pipe; the other two are removed with a screwdriver.

Lift the water housing free of the choke body as shown. The housing carries the bimetallic coil that operates the choke. If the coil is broken or deformed, replace it. Remove the white plastic gasket on the choke body.

The water housing can be further disassembled by removing the 11mm bolt that holds the assembly together. Unless the bimetallic coil requires replacement, there is usually no reason to disassemble the housing.

Three screws hold the choke body to the carburetor main casting. Remove them to release the choke body from the carburetor. Now, lift the choke body free. Take note of how the components inside the body are arranged.

If the choke is operative, don't dissassemble it. Chokes are complicated and have plenty of parts to get out of order or lose. Usually, all that is needed to service them is to thoroughly clean the linkage mechanism and check for tears in diaphraphms or gaskets. Be sure the cleaner you use is compatible with the choke's plastic parts. I've included the choke disassembly procedure for repair purposes.

Choke Disassembly—Free the gasket that seals the vacuum path into the body. Replace it if brittle or broken. Use an 8mm wrench to loosen the bolt holding the choke shaft in the choke body. Remove, in order: bolt, washer, control lever assembly, spring, spring locating cover, and slotted shaft washer. Remove the choke shaft from the choke body.

Loosen the screw holding the adjusting lever to the choke body. Remove, in order: screw, bushing, wave washer, adjusting lever with adjusting screw still attached (there's no need to remove it), and washer.

Three screws hold the vacuum diaphragm to the choke body. Remove them and catch the spring that pushes against the diaphragm. Separate the diaphragm from its mating surface and pull its rod from the choke body. Inspect the diaphragm for tears and replace if it has any.

Throttle Valves & Shafts—Don't remove the throttle valves and shafts unless **absolutely necessary**. The chances of making errors and ruining parts are high. So, what's absolutely necessary? If a valve is binding and can't be freed with cleaner, or the shaft is bent or worn, this should be taken care of. A worn throttle

Remove accelerator-circuit blanking screw to clean drilled passage. Use aerosol carburetor cleaner to assure passage is clear.

Auxiliary venturi removal: Unit should slip out easily. If stuck, use wood dowel from below to remove. Trying to lift with pliers will usually break venturi.

Remove choke water chamber. Establish index mark so chamber is not rotated incorrectly on reassembly.

Bimetallic coil that controls choke operation is heated by water chamber to slow choke action.

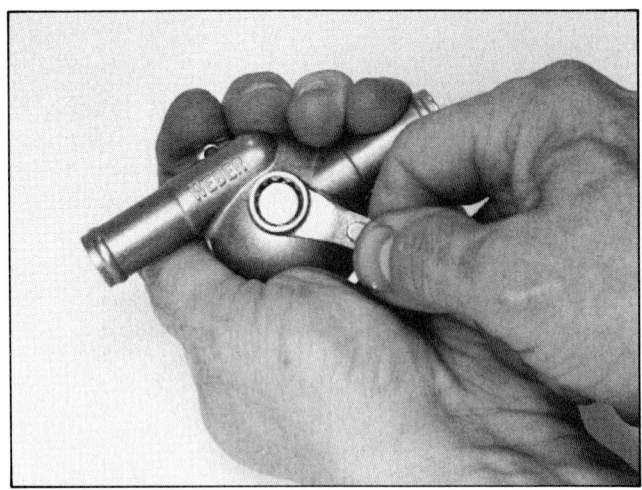
Water chamber comes apart easily by removing 11mm bolt. Unless it leaks, leave chamber assembled.

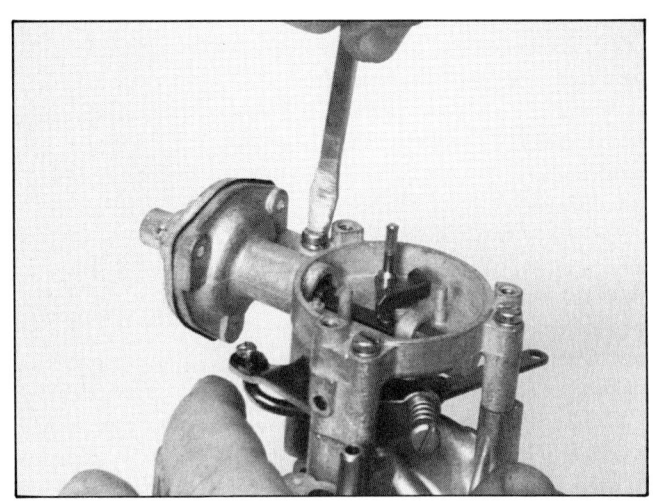
Remove choke body from carburetor. Three screws attach it through a stand-off to provide room for operating linkage behind housing.

How it all goes together: if you disassemble choke mechanism, reassembling can be difficult. Don't take it apart unless something is wrong with choke actuation.

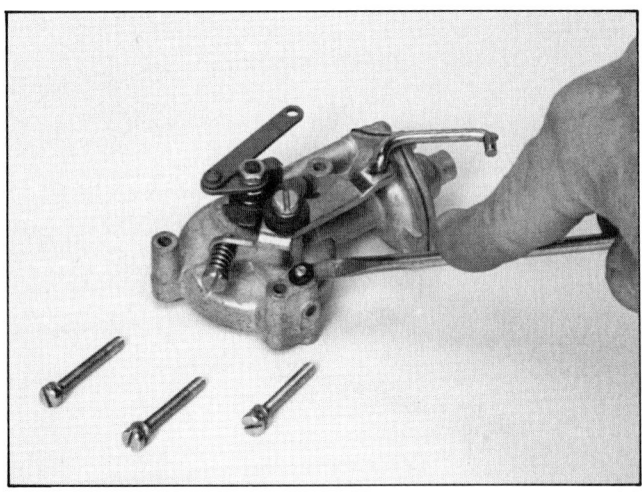
Remove rubber O-ring from back of choke housing. Ring makes an air-tight seal so manifold vacuum, operating on diaphragm, pulls off choke.

shaft will produce an erratic idle or an inconsistent return to a stable idle. If faced with these conditions, then seriously consider replacing the complete unit or having a Weber specialist do the repair. The removal and assembly procedures are included here for those with a stout heart, ample dexterity, and plenty of time.

Use a 12mm wrench to loosen the nut at the end of the throttle shaft. Don't put any pressure on the throttle valves. Use a screwdriver against the bellcrank assembly to keep the valves open while loosening the nut.

The shape of the parts that you'll remove from the throttle shaft will change, depending on the application. Generally, the next step is to strip the shaft of all its components. Lay them out carefully so you can reassemble them exactly the way they came off. Make a sketch to aid your memory. Begin by removing the nut, lock-tab washer, main arm, washer, lever, and shoulder washer. Note that the shoulder faces the carburetor body.

Release the fast-idle rod from the choke if the choke hasn't yet been removed from the carburetor. Use needle-nose pliers as shown to remove the split pin, see page 66.

Now remove the remaining lever and spring behind it. You may remove the nylon seal from the throttle shaft, but it's better to leave it in the cavity of the housing for now.

Now, remove the parts from the secondary throttle shaft: nut, lockwasher, washer, arm, and spring.

The next step is to remove the four screws that hold the throttle valves to the two shafts. As in the choke valve removal, the ends of the screws are punched to keep them from vibrating loose. Only slight pressure on a screwdriver can be applied to the screws to avoid bending the throttle shafts.

If slight pressure won't remove the screws, try unscrewing them in increments of about 1/2 turn, followed by screwing them back in, and then out again in 1/2-turn increments. If that method doesn't work, grind their ends so they can be removed without stripping the threads in the throttle shaft. The shaft itself is quite hard, so even if you do strip the screw, there will be enough thread left in the shaft to chase with a 4 × 0.7mm tap for new screws.

Mark both throttle valves so they can be replaced exactly as they were removed, and then remove the valves. Check to see that there are no burrs on the throttle shaft caused by removing the screws. Remove any burrs with a file, then pull the two shafts from the carburetor body. A spring will come out with the shaft. The spring is used to seat the nylon shaft seals in the carburetor body. Remove any nylon shaft seals that may hang up in the cavity of the carburetor body.

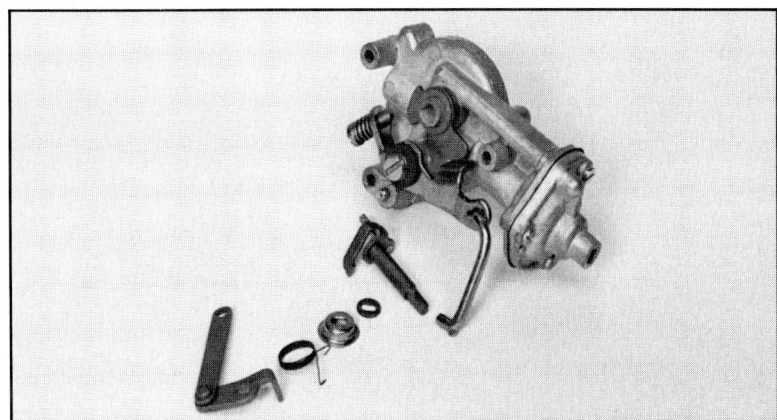

Make sketch so you understand assembly order of choke-shaft components before removing them. Photo gives assembly order, but not orientation of lever around shaft.

Adjusting-lever components laid next to choke-shaft components: Lever's orientation around shaft is easily forgotten and should be recorded.

Choke mechanism completely disassembled. Diaphragm, spring and cover are attached with three bevel-head screws. Diaphragm must be supple; replace if it isn't.

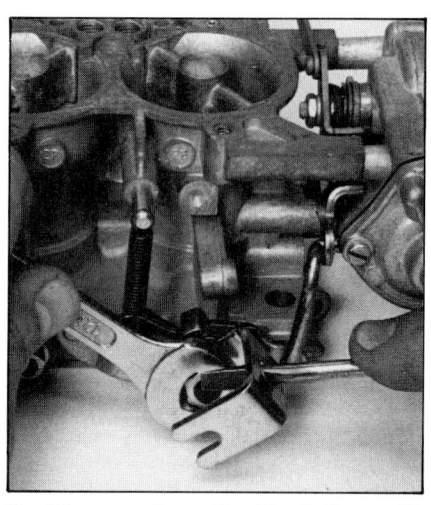

Don't loosen primary throttle-shaft nut with only a wrench. Use screwdriver to relieve any pressure against throttle valves. Note how levers are oriented before beginning disassembly.

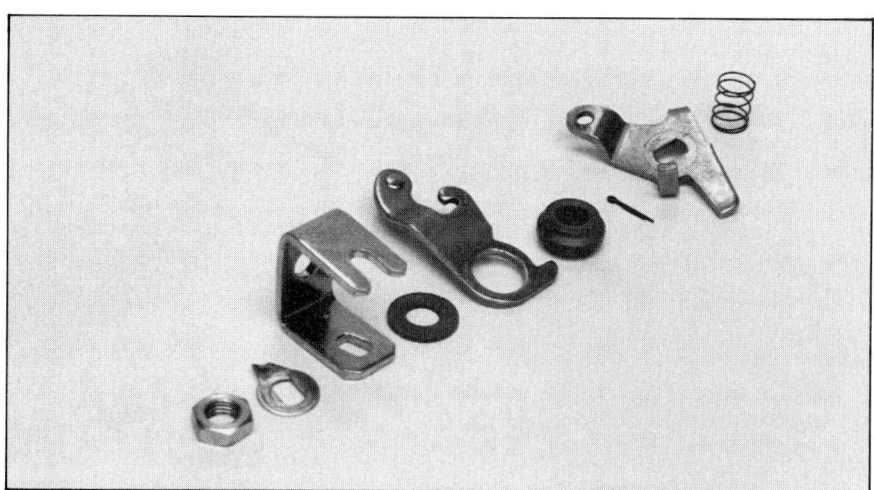

Throttle-shaft components removed in order: Orientation of levers around shaft must be noted before disassembly.

Remove choke's fast-idle rod from opposite end of throttle shaft. Cotter pin is small, somewhat hard to get at.

Secondary throttle-shaft components disassembled in order: Orientation of secondary throttle arm should be noted on a sketch before disassembly.

Removing screws from throttle shaft is a major step. Chance of success is good, but complications are unnumbered and unpredictable. Don't do this unless absolutely necessary.

66

If throttle-shaft screws are typical, this step is necessary. Grind off just enough of staked end to relieve upset threads. Use new screws at reassembly.

Throttle valves removed: Note white markings on valves for relocating at reassembly.

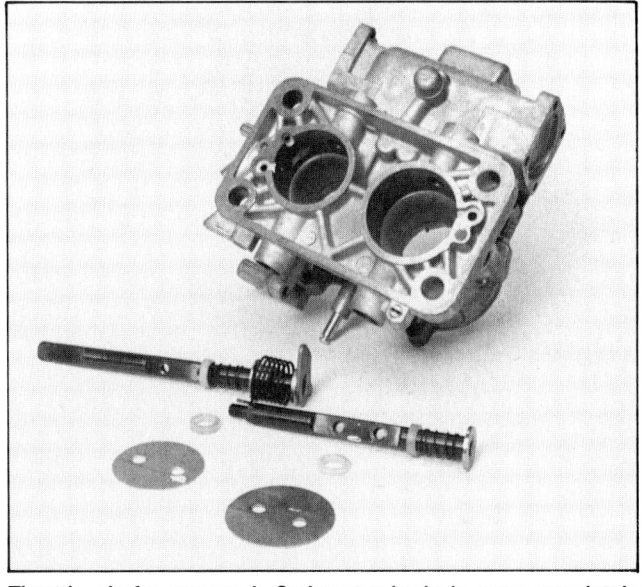

Throttle shafts removed: Carburetor body is now completely stripped. Note white plastic shaft seals on throttle shafts and their seating springs.

INSPECTION

Cleaning—See the Cleaning section, page 51, of the IMPE part of this chapter for tips on cleaning the carburetor.

Float Bowl Cover—The gasket that seals the float bowl cover to the carburetor main casting should show that the sealing surfaces of the two castings are parallel. Check the float bowl cover for broken flanges, cracks or other physical damage. Check the threads for the fuel filter plug and screw threads for the power valve.

Check the fuel filter for any damage to the screen. Blow it clean with compressed air to remove any dirt.

The choke shaft should be straight, with no visible wear where it bears on the float bowl cover. The choke valves should have smooth edges and be flat.

The floats should not show any evidence of leaking. Shake the assembly near your ear and listen for any gasoline sloshing inside. If the floats leaked, replace them.

The float needle valve and seat should not show uneven wear at their mating surfaces.

The diaphragm on the power valve assembly should not be damaged, nor

67

Staking throttle-shaft screws requires that you support throttle shaft. End of screw presents itself at an angle, so a dead-center punch is impossible. Note hammer used weighs less than five pounds. Shaft support is drive side of 1/4-in. drive socket, or short section of solid metal rod.

should its long spring be broken or corroded. Renew it as required.

Jets—Blow compressed air through the main jets, the air-correction jets, emulsion tubes and accelerator-pump jet to clean them. Never use a wire to probe or clean a jet; it will disturb the flow characteristics. Blow air through both sides of the accelerator-pump delivery jet. It is a one-way valve, so you should be able to blow through it only one way.

Press down on the center needle of the full-power valve to make sure it moves freely against its return spring.

If you removed the auxiliary venturis, blow though the passage and remove any burrs or dirt on their surfaces.

Main Casting—Check for any damage to the main casting. Use a straightedge to check the base of the carburetor for excessive warpage.

Remove the idle jets from their holders and check each jet by holding it up to a light. Blow through the jet if it's clogged. Check the diaphragm of the accelerator pump for damage.

The automatic choke assembly contains a lot of levers. Check that all operate freely. The bimetallic spring should not be bent or broken. Check the water chamber for signs of leakage.

Throttle Shafts—The throttle shafts should be straight, with no visible wear where they bear on the nylon seals. Double-check the threads on the end of the primary throttle shaft and the threads on both shafts where the valves are attached. The throttle valves should have smooth edges and be flat.

The springs that slide onto the throttle shafts should be strong and not deformed. Replace the nylon shaft seals whether or not they show any evidence of wear.

REASSEMBLY

The disassembly of the DGAV is, in itself, a lesson in reassembly. After all the components have been inspected and damaged parts replaced, begin reassembling the major components first. The disassembly photos will be helpful during assembly. Start with either the float bowl cover or main casting. The following notes give the most important points to remember in reassembly.

Throttle Shafts—There are two dangers in reassembling the throttle valves: bending the throttle shafts and not positioning the valves correctly. The first danger requires only care; the second, some strategy.

The holes in the valves are oversize so they can be positioned for a perfect seat in the bores. The auxiliary venturis should be removed to give free access to the attaching screws for this operation.

Install the throttle shafts with all their linkages and springs in place, then tighten the end nuts before installing the valves. This strategy assures that the valves are not moved while you're installing anything else on the throttle shafts. On the primary shaft, working outward from the innermost part, the order is:

Nylon washer
Spring
Arm
Shoulder washer, shoulder side to carburetor casting
Arm
Washer
Throttle bellcrank
Lock tab
Nut

On the secondary shaft, working outward, the order is:

Nylon washer
Spring
Arm—Engage arm with slotted arm on the primary shaft before slipping it onto the secondary shaft.
Washer
Lock washer
Nut

Let the valves position themselves in the bore and then tighten the attaching screws lightly, being careful not to move either the shafts or valves.

Double-check your work by holding the carburetor up to a bright light and looking around the circumference of the valve. You should be able to achieve an almost light-tight seal between the valve to the bore. If there is a crescent of light, it should be concentric with the valve.

Tighten the screws securely and recheck your work. The valves are held in place by the tightness of the screws. But the ends of the screws should be deformed (staked) to ensure they don't vibrate loose and drop into the engine.

Take the appropriate steps to prevent this catastrophe. Use fuel-resistant Loctite on the threads and stake the screw ends. Do this with a punch while supporting each so no force is transmitted to the throttle shafts.

Set a 1/4-in. drive, 1/2-in. socket on a steel or concrete surface with the square drive end facing up. Place the carburetor on top of the socket so a throttle valve screw head rests against the smooth surface of the socket. Then, place a punch down the venturi and position it on the end of the screw being supported by the socket. The valves won't allow the throttle shaft to rotate so you can hit the screw dead-center, but the angle is so small that it shouldn't matter. Try to deform the screw with a single sharp blow directed as closely centered on, and aligned to the screw, as possible.

You can also secure the screws with Vise-Grip pliers. Reach inside the venturi with a small pair of them to squish the threads on the exposed screw end. This will deform the end and keep the screw from backing out.

Carburetor Body—Place a rubber O-ring on both idle jet holders and screw them in place. Install the idle speed adjusting screw and set it so the throttle valve just begins to open. Run in the idle mixture screw all the way, gently so you don't damage the seat. Then back it out the number of turns that match its original setting. Two turns out is a good starting point. Screw in the accelerator-pump blanking screw.

Place the spring into the accelerator pump cavity and then cover it with the pump diaphragm; its nipple points outward. Be sure that the long arm on the accelerator pump cover fits against the underside of the cam on the throttle shaft. Gradually tighten the four shoulder screws that hold the pump to the carburetor, working in diagonal steps so the cover seats evenly.

Automatic Choke—Assemble the parts in the automatic choke housing. Begin by slipping the arm attached to the vacuum diaphragm into the housing, then seat the diaphragm. It's a one-way fit. Assemble its spring and cover. Tighten the three shoulder screws that hold the cover to the housing. Check to see that the Teflon bushing for the operating shaft is in place, then slip the operating shaft through the housing. The rounded end of the arm on the shaft should swing in the milled cutout of the vacuum diaphragm arm. Make sure the levers are returned to their original positions as shown.

Assemble the choke operating arm to the operating shaft. Work out from the casting: slotted washer, stamped cover, spring, arm, lock washer and nut. Don't bend the arm on the end of the shaft as you tighten the nut. Then, assemble the fast-idle operating arm working out from the casting: washer, arm, spring washer, locating bushing, screw. The lever and arm should point away from the vacuum diaphragm. Slip the heavy wire arm into the keyway on the fast-idle operating arm.

Fit the O-ring to the vacuum line and position the two arms. The flat-steel arm goes up to the choke butterfly valve and the heavy wire arm goes down to the linkage on the primary throttle shaft, where it's secured by a cotter-pin. Check to see that the adjusting screw clears its stop when the housing is placed against the carburetor. With everything in place, attach the choke housing to the carburetor with three screws.

Put the white plastic gasket in place over the housing. There's a locating pin so the gasket goes only one way; cover it with the water housing assembly. Make certain the loop in the end of the bimetallic coil fits onto the shaft projecting through the white plastic gasket.

Match up the index marks on the choke housing and water housing, then tighten the three screws that hold them together.

Jets—Replace the main jets, emulsion tubes and air-correction jets, auxiliary venturis, accelerator-pump delivery valve and power valve with its copper washer.

Float Bowl Cover—Fit the two dust seals into the float bowl cover. Install the choke shaft and valves, if they were removed. Use the same procedure and precautions as for the throttle valves. Next, attach the power valve assembly to the cover. Replace the float needle-valve seat if it was removed.

Slip the float needle valve onto the float assembly, place the valve into its seat and replace the pin that holds the floats to the top of the carburetor. Check to see that the needle moves freely in its seat.

Set Float Level—Whenever a DGAV is disassembled for repair, the float level must always be checked and set. Check float level by measuring the distance from the float's bottom edge to the bottom surface of the float bowl cover without the gasket installed. The float is in the correct position for measuring when the float tab is just touching, but not depressing, the spring-loaded ball in the needle valve.

This measurement is most easily made by holding the cover vertically with one hand so the float pivot is at the top and the float dangles straight down. See the accompanying illustration for the proper orientation and dimensions, page 70.

Move the cover slightly so that the float tab just touches the float needle assembly and then measure the distance between the float's bottom and cover's face using your other hand. For metal floats assembled from two halves, measure to the surface of the float itself, and not to the raised soldered joint at the circumference of the float.

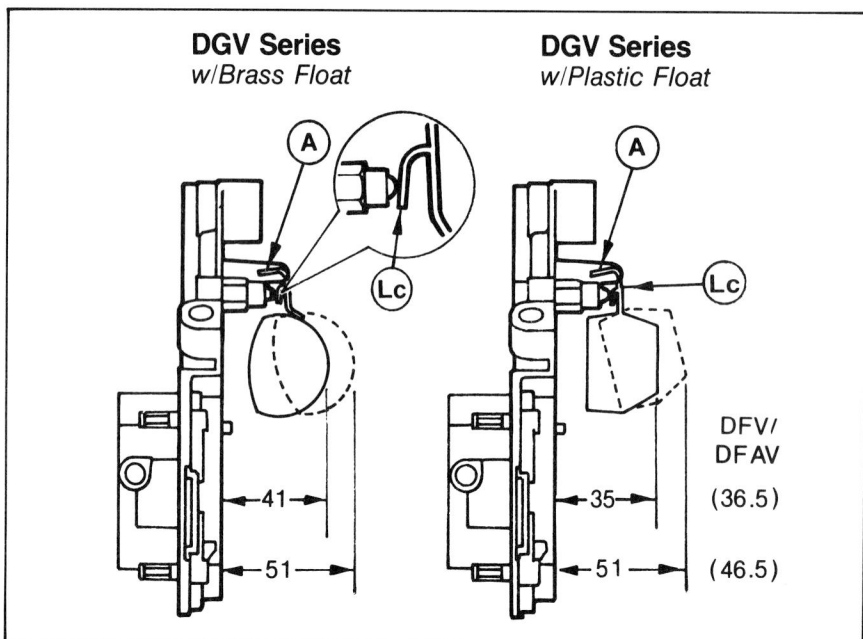

Float-level dimensions (in mm) for DGV series: For brass floats, use drawing at left; plastic floats at right.

Correct float level by bending the float arm tab that presses against the float needle assembly. Attach the float bowl cover to the main casting and secure all fittings; snug but not stripped.

EMISSION-CONTROLLED CARBURETOR: 32/34 DFT

The DFT is representative of Weber's response to exhaust emission regulations. More and more, manufacturers integrate emission controls so that the relationship between the carburetor and the rest of the engine is intimate. Gone are the days when you could just bolt on any carburetor that fits the mounting flange. Now, all carburetor manufacturers must contend with evaporative emission controls, exhaust gas recirculation, timed vacuum sources for the ignition and very subtle changes in air/fuel ratios. Failure to accommodate to these conditions could spell the end of the U.S. market for an independent carburetor manufacturer.

The DFT is an effort by Weber to continue providing aftermarket carburetors that improve performance, but keep all the emission controls of the original unit. Like the DGAV carburetor, it has two downdraft venturis and parallel throttle shafts with a progressive (beginning at about 2/3 throttle) linkage to the secondary throttle. Also like the DGAV, the DFT has Teflon bushings fitted to the throttle shafts to reduce shaft wear. That is the most common place for wear on those Webers not using ball bearings on the shafts.

This carburetor has several vacuum ports for operating engine and emission-control devices, depending on the requirements of the specific engine. Moreover, there are plenty of blank passages cast into the DFT body to permit additional vacuum taps to be added as needed. In this way, the basic DFT casting can be used to accommodate the needs of a variety of cars.

The main casting includes the two venturis and carries all the jets and their drilled passages, as well as the accelerator pump, power valve, fuel cut-off solenoid and numerous vacuum taps. The float bowl casting carries the choke mechanism and fuel-return check valve.

DGAV INITIAL SETUP & TUNING
Cold Engine:
• If float level has not been set, do it now.
• Check for fuel leaks. If using an electric fuel pump, turn on ignition so it operates and then check all fuel fittings for leaks. If mechanical pump is used, disconnect high-tension (large) wire from coil, crank engine for several seconds and then check for fuel leaks. Reconnect coil wire.
• Remove air cleaner. Open throttle all the way and then close it. The choke butterfly valve should close completely on a cold engine. Move choke-valve linkage to verify there is no binding. If there is, carefully free linkages so operation of the choke is correct. Be careful, if you bend a link, not to upset choke setting.
• Hold choke valve open and locate accelerator-pump outlet in primary venturi. Open throttle, then close it. Fuel should flow from accelerator-pump outlet. If it does not, correct problem.
• Fully depress accelerator pedal. Verify throttle plates fully open. If not, readjust linkage so they fully open.
• Verify, if necessary, that idle-speed and idle-mixture adjustment screws are at their initial settings. Connect tachometer to engine to monitor idle speed. Start engine and let it warm to operating temperature. This may require resetting idle speed so engine continues to run.

Warm Engine:
• Adjust idle speed to approximately 850 rpm. Any speed below 1000 rpm is acceptable so long as intermediate or main circuits are not operating and engine continues to run.
• Adjust idle-mixture screw so engine idles at maximum speed. Start by turning screw out in equal increments of 1/2 turn—rotate slot in screw's head exactly 180°—until engine rpm begins to decrease. If turning screw out only increases engine rpm, reset screw to the initial setting and readjust baseline idle speed so it is as low as possible. Then, repeat idle-mixture adjustment. As engine rpm decreases, turn in idle-mixture screw until maximum idle speed is obtained.
• Readjust idle speed to 850 rpm, or as specified in owner's manual or underhood label.
• Disconnect tachometer.

EMISSION-CONTROL PARTS

The DFT can have the following emission-control devices.

Electrically Heated Choke—An electrically heated choke sets an enriched mixture independent of the actual operating temperature of the engine. The electrical heater (operated by current from the alternator) controls choke operation over a shorter period of time and more precisely than a water-heated automatic choke. A vacuum pull-off opens the choke plate under high manifold vacuum to keep from creating an excessively over-rich fuel mixture when the car is decelerating with a closed throttle.

Fuel Cut-off Solenoid—When the ignition is switched off, fuel is shut off from the idle circuit. The solenoid valve prevents *run-on* or dieseling. Check its operation by turning on the ignition to energize the circuit, but don't start the car. Pull the electrical lead off the solenoid valve and reattach it. You should hear the valve click as it opens and closes when the current is interrupted.

Float-Chamber Vent Valve—This valve directs vapors from the float chamber to a carbon canister when the engine is not running. Hydrocarbons evaporating from the float bowl are stored (engine off) in the canister and later purged (engine running).

Fuel-Return Check Valve—This check valve is part of the fuel delivery circuit in those applications that return excess fuel to the tank. It assures that fuel flows only one way in the system.

Sealed Idle Mixture Screw—The mixture adjustment screw may be sealed with a plug in some applications. This is to prevent unwarranted adjustment of the idle mixture. Emission-controlled cars run marginally lean, so increasing idle mixture richness is a tempting way to improve driveability.

Power Valve—At near WOT, the power valve momentarily increases fuel richness for improved performance.

DISASSEMBLY

See the sidebar, page 45, for information on special Weber tools. These tools

DFT is example of carburetor designed for emission control.

are not needed for most home-bound repair procedures. Beyond a thorough cleaning, setting the float level and replacing rubber parts and gaskets, complete Weber overhauls are better left to specialists.

It's not practical to disassemble this carburetor completely just to fix one part. You can do more accidental harm disassembling it than could ever result from normal wear. A little extra time spent diagnosing the problem can save the grief of breaking a perfectly good part during disassembly.

The most common wear points on a DFT are the throttle shaft bearing surfaces on the carburetor body. The DFT has nylon seals around the shafts where they pass through the main casting. Replacing these seals means removing the throttle plates and shafts, the most challenging repair operation for the carburetor. If the throttle shafts are not worn, a DFT can probably be repaired on the car. If they are worn, consider replacing the complete unit. The cost and effort could be less than throttle shaft renewal.

The following procedure covers the complete disassembly of the carburetor when it's off the car:

- Removal and disassembly of float bowl cover.
- Removal of components accessible when float bowl cover is removed.
- Disassembly of components on outside of carburetor.
- Removal of throttle shafts.

There is no best order in which assemblies can be removed. Think of the carburetor as a collection of modules, any one of which can be worked on without disturbing the others. Here are the three main modules of the DFT:

- Float bowl cover:
  Carries choke mechanism, float, float needle valve, fuel filter and check valve
  Gives access to all fuel jets, air-correction jets and emulsion tubes, accelerator-pump delivery valve
- Exterior of carburetor:
  Idle jets
  Accelerator pump
  Automatic choke assembly
  Power valve
  Fuel cut-off solenoid

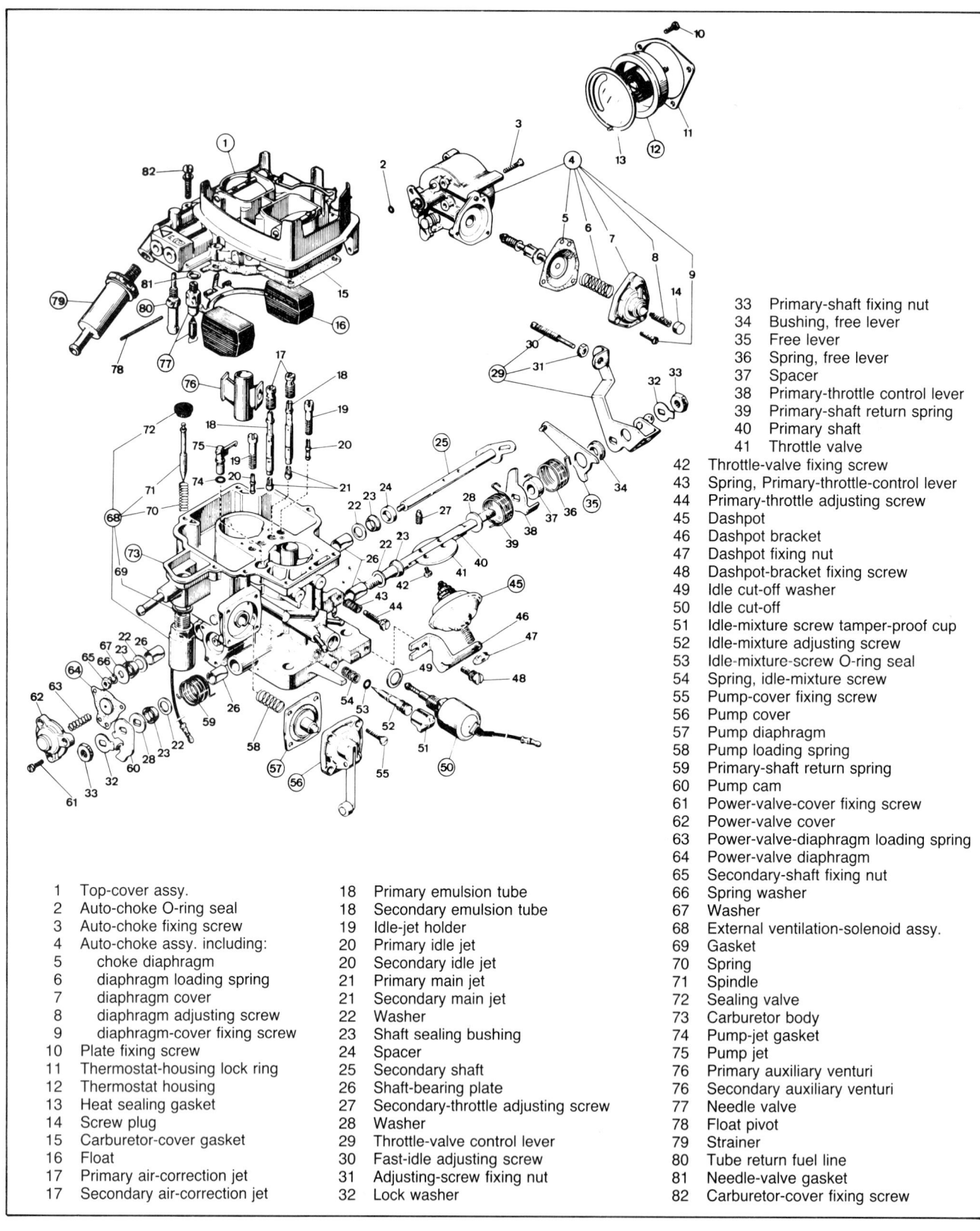

32 DFTA: Note (45) dashpot, (50) idle cut-off solenoid, and (68) float-bowl-vent solenoid. Drawing courtesy Redline, Inc.

Remove DFT float-bowl cover. Plastic, not metal floats are typical of modern emission-controlled Webers. Note that choke assembly stays attached to float-bowl cover.

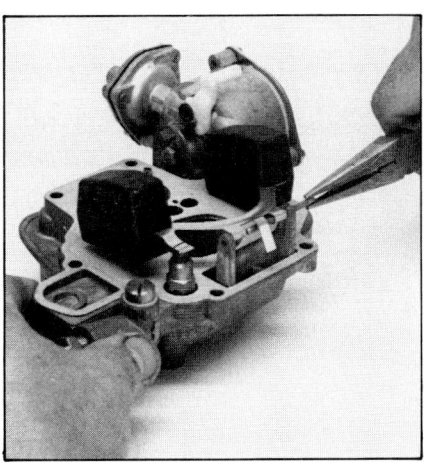
Remove plastic float assembly. Float-bowl-cover gasket is trapped under floats. Needle valve that hangs from float assembly by thin wire is easily dropped/lost. Note incorrect method to remove float-pivot pin—extract through solid boss, not split one.

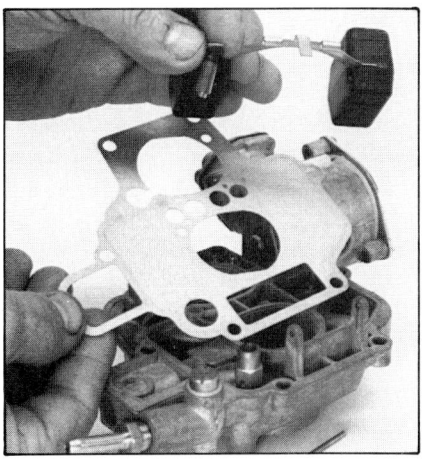
Gasket removal: Clean, untorn gaskets can be reused, so handle with care.

- Throttle shafts:
  Gives access to shaft seals and throttle plates

Therefore, the procedure that follows only suggests the order in which assemblies may be removed. A wide variation is possible, depending on the repair needs of the carburetor.

If you're going to do a complete overhaul, buy an overhaul kit. If you're doing anything less than a complete overhaul, you may be able to save most of the old gaskets and diaphragms necessary to reassemble the carburetor successfully. If you're trying to save a gasket or diaphragm, work very slowly and carefully to separate it from its mating surface.

Begin the overhaul by cleaning the outside of the carburetor as completely as possible. Scrape baked-on dirt with a knife or screwdriver, but be careful not to gouge out pieces of the carburetor in an attempt to clean it. Use an aerosol can of carburetor cleaner to remove varnish and any remaining dirt. Follow the precautions and instructions on the cleaner's can.

Float Bowl Cover—Remove the fuel filter by unscrewing it from the float bowl cover casting. Note: Not all DFT carburetors have a filter. Remove the six float bowl cover screws and lock washers. Open the throttle so the fast idle adjusting screw is free of the choke housing. Now lift off the float bowl cover. Both the float bowl cover gasket and choke assembly will come off with the cover.

If you can't remove the cover with a slight pull, turn over the carburetor and use a screwdriver handle to tap on the choke assembly to loosen the cover from the carburetor body. You can also tap lightly on the fuel filter extension.

If you're trying to save the gasket for the float bowl cover, slide a knife along the gasket, separating it from the mating surface as slowly and carefully as possible. The paper gasket may be brittle from age and heat. It's not always possible to remove one whole. If you tear the gasket, replace it.

Use needle-nose pliers to pull the pin that holds the float assembly to the cover. Or take a suitable punch and tap out the pin. Don't pull or push the pin through the split boss; extract it through the solid one. The float assembly will lift off, taking the needle valve with it. The needle valve is attached to the float assembly with a spring wire loop. Don't bend the wire as you slide it from the float. The gasket is trapped onto the cover by the float assembly. When you have removed the float, you can carefully lift off the gasket.

If necessary, remove the needle-valve seat. It has an aluminum washer that may stay attached to the float bowl cover. Be sure to replace the seat with only one washer, otherwise float level adjustment will be off.

Remove the fuel-return check valve if it is fitted.

Automatic Choke—If the choke is working, *don't* disassemble it. Chokes are complicated and have plenty of parts to get out of order or lose. Usually, all that is needed to service them is to thoroughly clean the linkage mechanism and check for tears in diaphragms or gaskets. Be sure the cleaner you use is compatible with the choke's plastic parts. I include

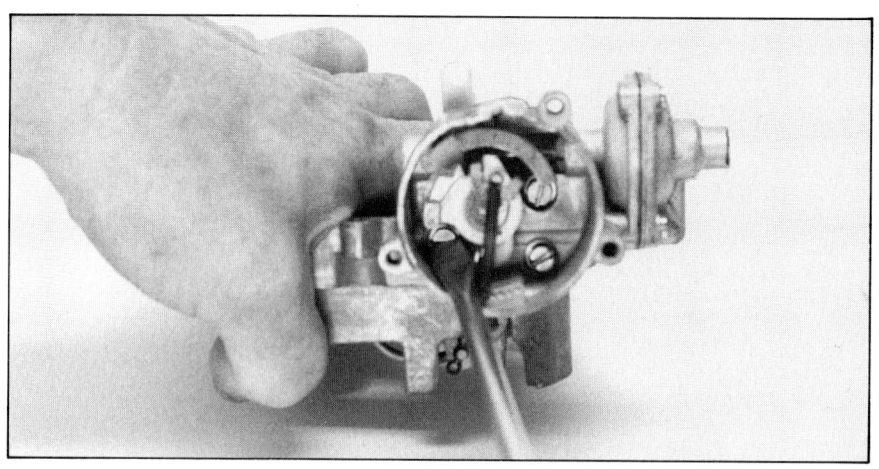
Remove one of three screws attaching choke assembly to float-bowl cover. Screw being loosened is normally covered by choke mechanism, which must be rotated to reach screw. Be careful not to stretch spring.

Choke body removed from float-bowl cover: Unless choke is broken, don't disassemble any further.

Choke pull-off diaphragm cover removed to show diaphragm, spring.

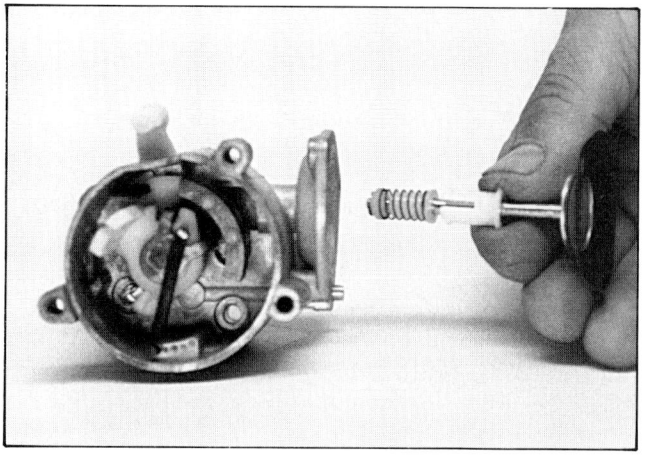
White plastic snap-in collar secures diaphragm assembly in choke body. Use needle-nose pliers to remove.

the disassembly procedure here for repair purposes.

The operation of the automatic choke is adjusted by rotating the electric heater assembly in relation to the choke body. The three screws you're about to loosen permit changing this adjustment. The factory punches index marks to calibrate the proper relationship between the two parts. There are three punch marks on the main choke casting and one on the heater. The middle punch mark is the normal position for the choke. The one to the left when facing the choke housing is for a richer setting; one to the right, leaner.

Find the punch marks and make a note of them so you can reassemble the halves in the same relationship. Make your own reference mark if there's any question about the relationship of the parts.

Now, remove the three screws holding the automatic choke bimetallic coil housing to the choke body and remove the housing. The housing carries the bimetallic coil that operates the choke. If the coil is broken or deformed, replace it. Remove the white plastic gasket on the choke body.

Three screws hold the choke body to the carburetor main casting. One screw is hidden behind the large white plastic segment that is part of the choke operating mechanism. Rotate it so you can get to the screw. Undo the three screws to remove the automatic choke assembly from the float bowl cover. Be careful not to bend or damage any of the choke mechanisms. Discard the rubber O-ring that fits between the choke assembly and the float bowl cover. The O-ring seals a vacuum line and should be routinely replaced.

Lever assembly removed from choke body: Removal of rear holding nut requires steadying shaft with needle-nose pliers. Tabs on lever assembly are easily bent, destroying choke operation.

Disassembled shaft shows tab with spring attached, and weight to control operation of over-center mechanism.

Lift the choke body free by unhooking the choke lever at the rear of the choke body from the main operating link. A staked tab on the main link holds the link to the choke lever. Unless the choke is broken, it's not necessary to disassemble it further.

Choke Disassembly—If it is necessary to disassemble the components inside the choke body, begin by taking a good look at the relationship of the components inside the body and how they are located. The choke is a collection of levers that don't reassemble self-evidently. Take time to sketch precisely how all the components fit together, updating your sketch as you remove each part. Pay attention to the exact relationship of parts that pivot on the main control shaft.

Remove the three shoulder screws that hold the choke pull-off diaphragm cover. There is a weak spring inside the cover that may fall out as the cover is removed.

The diaphragm itself is held in by a plunger assembly that is captured by a snap-in plastic collar. Use needle-nose pliers to compress the collar from inside the choke housing and pull the assembly free.

This step describes the removal of the lever assembly inside the choke body. First, remove the long tensioning spring from inside the casting. From the rear, use an 8mm wrench to loosen the nut that holds the choke shaft at the back of the choke body. Holding the shaft to remove the nut presents something of a problem. Use needle-nose pliers to hold the shaft from inside the choke body while you apply torque to the nut. Be careful not to bend the two long metal legs that rotate with the shaft inside the choke body.

Remove from the rear, the shaft nut, lock washer and choke lever, then withdraw the shaft and white plastic lever from inside the choke body. Next, still working from the rear, press out the steel bushing that holds the white plastic fast-idle cam assembly captive. The bushing diameter is 5.79mm (0.228 in.). A thrust washer on the bush fits between the fast idle cam assembly and the casting.

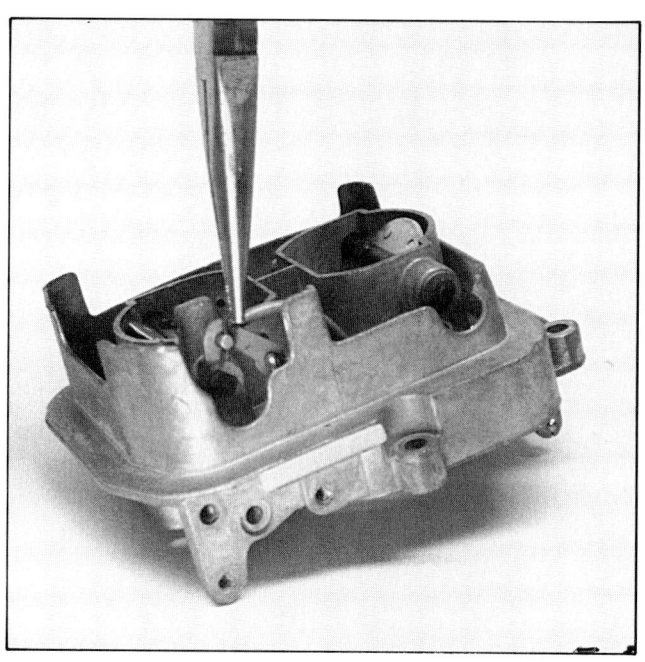
Remove R-clip to begin disassembly of choke valves. Clip is easily lost, so secure it.

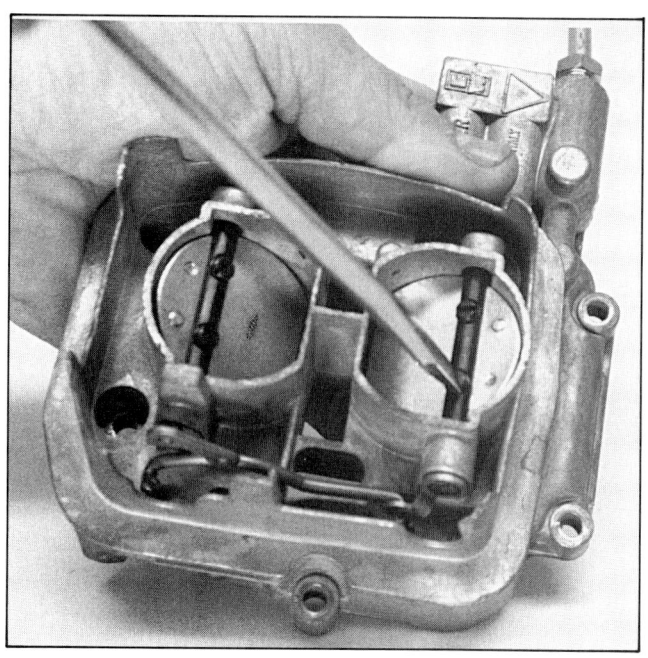
Choke valves remove just like throttle valves. Removal offers potential for disaster. Unless valves are bent or binding, leave them in place.

Choke Valves—Detach the two R-clips that hold the heavy-wire operating links to the choke valve lever at the primary venturi. Next, remove the main vertical operating shaft by pulling its staked end through the white plastic seal on the float bowl cover housing. Then, remove the white plastic plug and seal.

Mark each choke valve so it can be replaced exactly as it was removed. Then, very carefully unscrew the choke-valve attaching screws. The screws are staked so they can't vibrate free. You may have to grind off the ends of the screws to remove them without destroying the threads in the choke-plate shafts.

Withdraw the valves, then use a file to smooth any burrs on the shafts that could damage the Teflon bushing on the ends of the shafts. Remove the shafts and their Teflon bushings. Loosely reassemble the parts off the carburetor to retain their correct relationships.

Accelerator Pump—Remove the four accelerator-pump shoulder screws in increments, working around the cover. Then, remove the cover, a combination gasket/diaphragm and a spring. Don't try to separate the gasket and diaphragm.

Use a screwdriver to free the accelerator-pump delivery tube and jet assembly from the carburetor body. Then remove the O-ring seal holding the assembly in its bore. Blow through the assembly from each end to verify one-way flow.

Fuel Cut-off Solenoid—Remove the fuel cut-off solenoid valve. The flats on the solenoid valve may not be accessible with a wrench because of interference by some vacuum ports. You may be able to unscrew the solenoid by hand. If not, wrap some tape around the solenoid and carefully use pliers to start the valve out of its threads.

There is a plunger/valve assembly that operates freely inside the solenoid. Be careful not to drop the assembly out of the solenoid when you remove it from the carburetor. Also, you must replace the solenoid using the same number of gaskets, because the gaskets are actually shims that affect the seating of the valve.

Vent Solenoid—If fitted, remove the float-chamber vent solenoid, washer, rod, spring and seal.

Power Valve—Remove the three shoulder screws holding the power valve cover, then remove the cover, spring and diaphragm. Note that the diaphragm has two layers of rubberized fabric.

The power valve has a one-way check ball captured in a bronze bushing that is pressed into the main casting. Use a toothpick or other soft probe to be sure that the ball can be depressed and returns to its seat under spring pressure.

Dashpot—If fitted, remove the dashpot mounting screw and dashpot.

Idle Mixture Screw—The idle mixture richness adjustment screw is covered by a plastic cap. Pierce the cap with a small screwdriver and pry it out of its bore to get at the adjusting screw underneath. Count the number of turns to gently seat the screw, record it, and then back out the screw. Remove it with its spring, then remove the O-ring from the screw. If the

Accelerator-pump diaphragm has gasket attached, will probably tear if you try to separate the two.

Power-valve diaphragm removal: Check ball can be seen in center of housing that is part of main carburetor body.

Break through tamper-proof seal for access to idle-mixture screw.

Auxiliary venturi removal: Stuck venturis should be driven out from beneath using wood dowel.

O-ring becomes brittle or shrinks, the idle screw can rotate and change the mixture. Replace the O-ring to maintain a stable idle mixture.

Auxiliary Venturis—Remove both the primary and secondary auxiliary venturis by lifting them out. If either venturi sticks, it can be driven out with a wood dowel from the bottom of the carburetor.

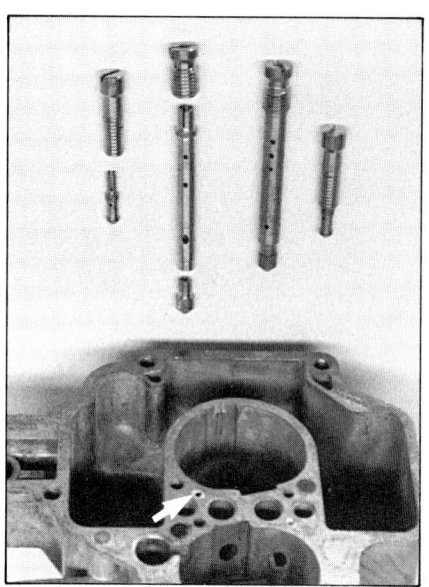

Jet set: primary and secondary main and idle assemblies. Left-hand set is pulled apart to show individual fuel jets and air-correction scheme. Idle jet has integral air-correction passage. Air brake jet itself is a bushing (arrow) visible on carburetor body.

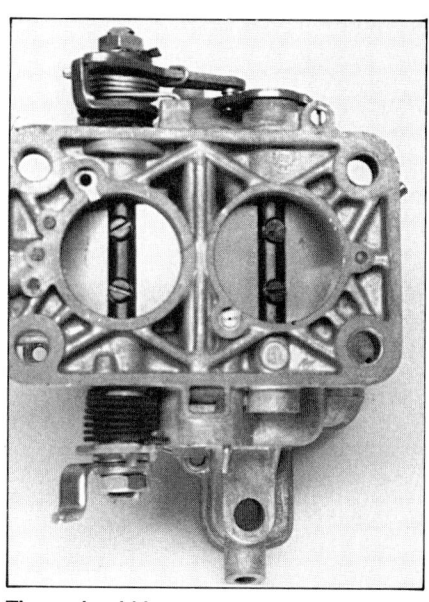

There should be an *overwhelming* reason to remove throttle-valve screws. Otherwise, leave them alone.

Jets—Unscrew the four assemblies containing the primary and secondary jets for each venturi. Pull them apart for inspection. The main jet assembly has three sections and the idle jet assembly has two.

The fuel jets are at the bottom end of each assembly and are quite hard to pull out. Be careful not to damage the jets while trying to separate them from their holders with pliers. Clean the jets with solvent or compressed air: never use a wire or drill to clear a passage.

Idle Speed Screw—Remove the idle speed adjustment screw and spring, if necessary.

Throttle Shafts—Don't remove the throttle valves and shafts unless **absolutely necessary**. The chances of making errors and ruining parts are high. So, what's absolutely necessary? If a valve binds and can't be freed with cleaner or very minor filing of its edge, or the shaft is bent or worn.

A worn throttle shaft produces an erratic idle speed or an inconsistent return to a stable idle. If faced with these conditions, seriously consider replacing the complete unit or have a Weber specialist do the repair. The removal and assembly procedures are included for those with a stout heart, ample dexterity, and plenty of time.

There are a lot of pieces on both ends of the DFT's primary throttle shaft. The secondary shaft is virtually one-piece. It is easy to confuse the order of reassembling the primary-shaft pieces unless you make your own clear notes on how they disassemble.

Begin dismantling the throttle shafts by bending back the tab washer and removing the 12mm nut at the end of the primary throttle shaft nearest the idle speed adjustment screw. It isn't necessary to remove the nut at the other end of the shaft. Don't put any pressure on the throttle plates. Use a screwdriver against the throttle lever to keep the throttle valves open while loosening the nut. Lay out the components in order as you remove them and draw a reference sketch or take a snapshot.

The approximate order of removal is: nut, tab washer, throttle lever, spacer, spring, spacer, stop lever, spring, washer. The actual order on the unit varies depending on the application, so make a list. The throttle lever can go on at least two ways. Your notes should be clear enough that you know its rotation exactly—how it is positioned as it moves around the shaft's axis—and which side goes outward.

Next, remove the four screws holding the throttle plates to the two shafts. As with the choke valve, the ends of the screws are staked to keep them from vibrating loose. Only slight torque can be applied to the screws with a screwdriver to avoid bending the throttle shafts. If slight torque won't remove the screws, grind off their ends so they can be removed without stripping the threads in the throttle shaft. The shaft is hard, so if you strip a screw, there should be enough thread left in the shaft to chase with a 4 × 0.7mm tap so new screws can be used.

Primary throttle-shaft components: Correct orientation of two levers is not represented here.

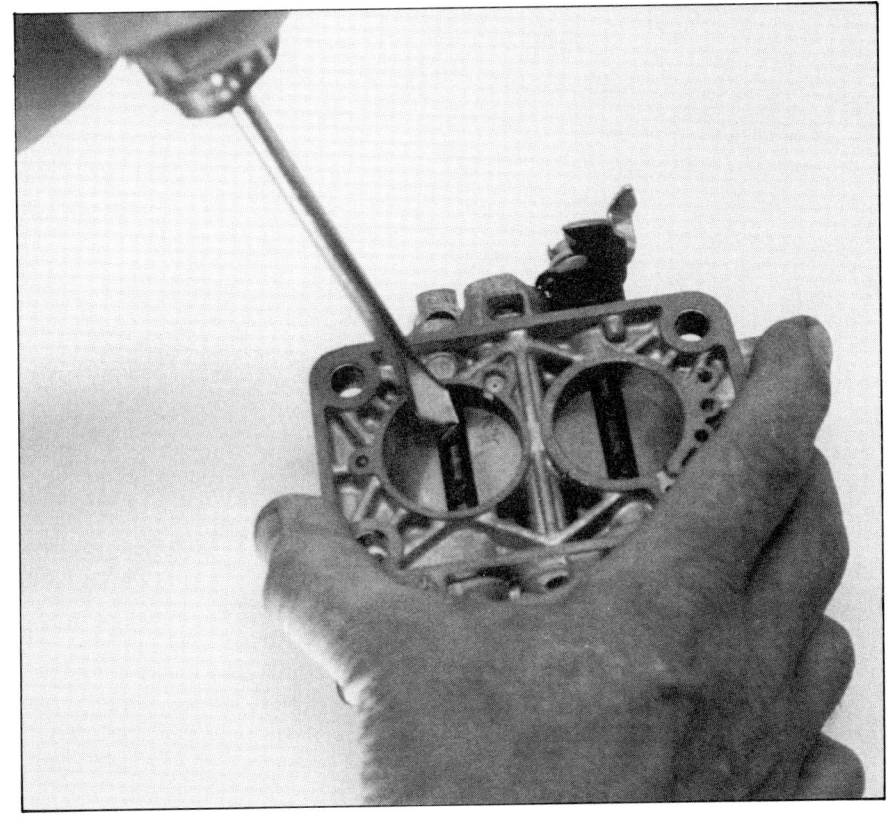

Remove throttle-shaft screws. Safest procedure is to grind end of screws to remove staked threads. If you don't, about one of four screws will strip throttle shaft threads, requiring replacement of throttle shaft, or repair of its threads.

Throttle-valve removal: Valve has been marked to show direction of idle-mixture screw so correct valve reassembly is ensured.

Remove burrs from throttle shaft before extracting it. Burrs damage shaft bearing surface in carburetor body, cause leaks and premature wear.

Primary throttle-shaft parts layout: Orientation of levers is not necessarily accurate here.

Mark both throttle valves so you can replace them exactly as they were removed, then remove the valves. Double-check that there are no burrs on the shaft to damage the Teflon seal when the shaft is withdrawn from the bores. File the shaft as necessary, but be careful not to damage the venturi with the sharp tip of the file. When you have determined that the shaft is smooth, grasp the accelerator-pump cam to pull the shaft from its bore. Remove the return spring and then the Teflon seals and washers.

If you wish to further disassemble the primary shaft, note the orientation of the accelerator-pump cam, and remove it from the shaft. You may also remove the nut, tab washer and spacer washer to strip the main throttle shaft completely.

It's a good idea to lay out all the shaft components in order and make a sketch, so you will not be confused during reassembly.

Before pulling out the secondary throttle shaft, check to see that it is free of burrs. File off any, but be careful not to nick the venturi. Then, remove the secondary throttle shaft with its spring, washer and spacer. Remove the Teflon bushings from the main casting.

INSPECTION

Cleaning—See the cleaning section, page 51, of the IMPE part of this chapter for tips on cleaning the carburetor.

Inspect all drilled passages in the carburetor as you clean it with a spray or an ear syringe full of solvent. It's essential that you wear goggles when doing this; work in a well-ventilated area.

The aerosol cleaner cans come with a plastic tube that inserts into the spray nozzle. Use the tube to probe the passages while spraying cleaner. It's unusual for passages to be blocked. Usually, dirt blocks the jets before the passages. A more common problem to look for is the loss of plugs used to block the ends of drilled passages.

Float Bowl Cover—The gasket sealing the float bowl cover to the carburetor main casting should show that the sealing surfaces of the two castings are parallel. Check the float bowl cover for broken flanges, cracks or other physical damage. Check the threads for the fuel filter plug and threads for the power valve.

Choke Shaft—It should be straight, with no visible wear where it rotates in the float bowl cover casting. The choke valves should have smooth edges and be flat.

Float—The floats should not show any evidence of absorbing fuel. Weigh the float assembly to see if it has absorbed fuel. If it has, replace it. Don't attempt to repair the assembly.

Needle Valve—The float needle valve and its seat should wear evenly where they seat against each other.

Power Valve—The diaphragm on the power valve assembly should not be damaged, and its spring should not be broken or corroded. Press down with a soft probe (toothpick) on the check ball for the power valve to make sure it moves freely against its return spring.

Jets—Blow through the main jets, air-correction jets, emulsion tubes and accelerator-pump jet to clean them. Never use a wire or drill to probe or clean the jets. Blow through the accelerator-pump delivery in both directions. It is a one-way valve, so you should only be able to blow through in one direction.

Remove the idle jets from their holders and check each jet opening by holding it up to a light. Blow through the jet if it appears to be clogged.

Auxiliary Venturis—If they have been removed, blow though the passage and remove any burrs or dirt from the surfaces of the venturi.

Main Casting—Check for any damage to the main casting. Use a straightedge to check the mating surface of the carburetor base for warpage.

Accelerator Pump—Check the diaphragm of the accelerator pump for damage. Replace it if punctured or cracked.

Automatic Choke—The automatic choke uses many levers. Check to see that they all operate freely. The bimetallic spring should not be bent or broken.

Throttle Shafts—The throttle shafts should be straight, with no visible wear where they rotate in the nylon seals. Roll them on a flat surface to check for straightness. Double-check the threads at the end of the primary throttle shaft, and those in both shafts where the throttle valves are attached. The throttle valves should have smooth edges and be flat.

All the springs that install on the throttle shafts should be strong and not deformed. Replace the Teflon shaft seals whether or not they show any evidence of wear.

REASSEMBLY

The disassembly of the DFT is the best lesson in reassembly. After all the components have been inspected and damaged parts replaced, reassemble the major components. The disassembly sequence photos should be helpful during assembly. Start with either the float bowl cover or the main casting. The notes below give the most important points to remember during reassembly.

Throttle Shafts—There are two dangers in reassembling the throttle valves: bending the throttle shafts and not positioning the valves correctly. The first danger requires only care; the second, some strategy.

The holes in the valves are oversize to permit them to be positioned for a perfect seat in the bores. Remove the auxiliary venturis to give free access to the attaching screws during this operation.

Begin reassembly by refitting the throttle shafts and replacing the Teflon bushings. Check to see that there are no small burrs on the shafts from the throttle plate screws. Smooth the shafts with a fine file.

Reassemble the shafts. Use the disassembly sketch or photo you made to help orient the parts. There are three return springs fitted to the primary shaft. The largest one goes on the end of the primary shaft that also has the accelerator-pump cam on it. The smallest goes next to the carburetor body on the other end of the primary shaft. Another heavy-wire return engages the secondary throttle-shaft bellcrank.

The secondary shaft is never really disassembled, because it has no springs or removable arms. Check the Teflon seals and bushings on the secondary before refitting the shaft.

Prepare the throttle plate attaching screws for reassembly by chasing them with a 4×0.07mm die and chasing the threads in the throttle shaft with a tap.

Throttle Valves—After the shafts are reassembled and installed, back off the idle adjustment screw so that the throttle plates can seat against their bores. Slip the throttle valves into their slots on the throttle shafts and make sure they seat uniformly against the throttle bore by snapping them closed. Hold the carburetor up against a bright light, and check for uniform light leakage around the circumference of the valves. Readjust them as necessary to achieve a uniform closing that is as complete as possible. Don't move the throttle shafts once the throttle plates are seated.

Install the screws into the throttle shafts and lightly tighten them to hold the plates. Then, operate the throttle shafts. Verify that the throttle plates are closing completely by looking at them against a bright light. Tighten the screws. The valves are held in place by the tightness of the screws, but the ends of the screws

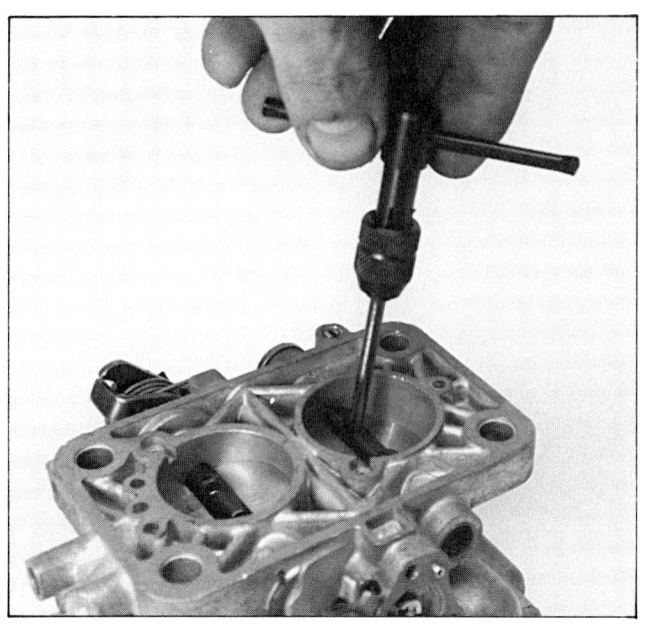
Chase throttle-shaft threads to assure best possible reassembly of throttle valves and screws.

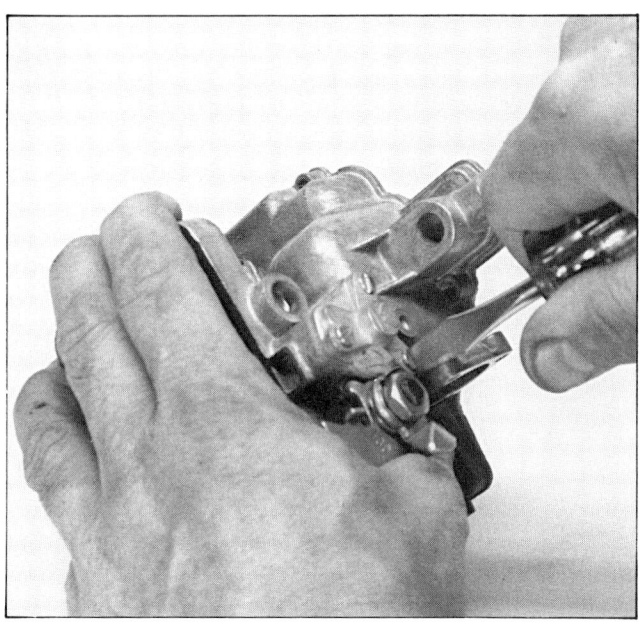
Reassemble power valve to carburetor body. Valve is one-way fit. Incremental tightening of screws avoids wrinkling diaphragm.

must be staked so they don't vibrate loose and drop into the engine.

Take the appropriate steps to prevent this catastrophe. Use fuel-resistant Loctite on the threads and deform the screw ends. You can deform the screw ends with a punch only if you support them correctly so no force is transmitted to the throttle shafts.

Set a 1/4-in. drive, 1/2-in. socket on a steel or concrete surface with the square drive end facing up. Place the carburetor over the socket so a throttle valve screw head rests against the smooth surface of the socket. Then, place a sharp punch down the venturi and position it on the screw that is supported by the socket.

The valves won't allow the throttle shaft to rotate so you can hit the screws dead-center, but the angle is so small that it shouldn't matter. Deform the screw with a single sharp blow directed as close to in-line with the screw as possible.

You can also deform the screw threads with Vise-Grip pliers. Reach inside the venturi with a small pair of them to squish the threads on the exposed screw end. This will deform the end and keep the screw from backing out.

Install the idle speed adjusting screw and set it so the throttle valve just begins to open.

Jets & Venturis—Reinstall the main and idle jet assemblies. The thread diameters are different, so there is no danger in confusing the two assemblies.

Reinstall the venturis. They are a one-way fit and should slide in easily.

Idle Mixture Screw—Check that the rubber O-ring is undamaged on the idle mixture richness needle screw. Replace it if required. Reinstall the screw, seat it lightly, then back it out 1-1/2 turns. Final adjustment of the screw is made on a warm engine using an HC/CO meter.

Power Valve—Inspect the diaphragm on the power valve and replace it if there is any evidence of hardening or tearing. Make sure the check valve in the carburetor body moves freely, then install the power valve in the carburetor. It attaches only one way, with its vacuum passage acting as a locating boss. Tighten the screws in small increments to ensure that the diaphragm doesn't wrinkle during installation.

Fuel Cut-off Solenoid—Refit the gasket(s) to the fuel cut-off solenoid. Hold the solenoid vertically so its plunger points up, then lower the carburetor body onto the solenoid, carefully seating the plunger into the carb body. Tighten it snugly by hand. Don't forget to connect the wire to the solenoid when refitting the carburetor to the manifold.

Accelerator Pump—Check the O-ring that seals the accelerator-pump delivery valve and replace it if deformed or brittle. Blow through the valve to check for one-way operation, and replace it if it's defective. Press the delivery valve into the carburetor body.

Check the accelerator-pump diaphragm and spring for deterioration. Replace either if defective. Fit the spring to the carburetor body and cover it with the diaphragm, then attach the cover to the carburetor body with four shoulder

Reassemble accelerator pump to main body. Like power valve, care should be used to avoid wrinkling diaphragm.

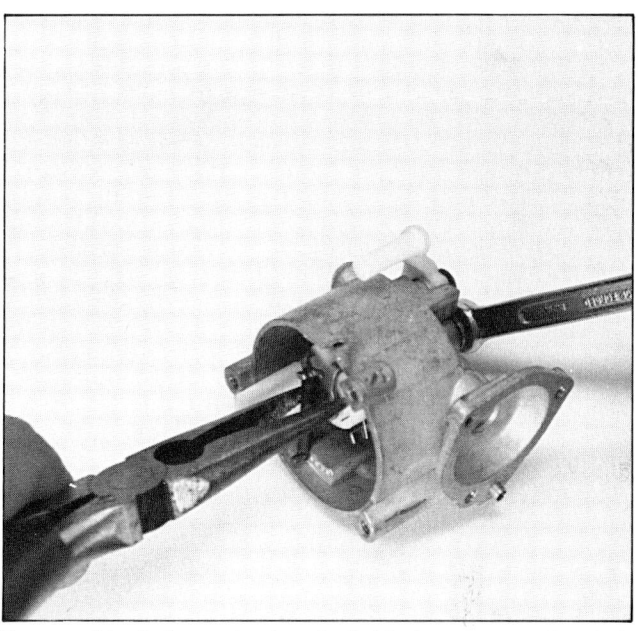
Reassemble choke mechanism. Probably the most complex challenge during reassembly, choke mechanism should be checked and double-checked every step of the way.

screws. Tighten the screws incrementally to ensure the diaphragm doesn't wrinkle.

Automatic Choke—Refit the choke shafts, then reassemble the choke plates to their shafts, using the same technique to stake the ends of the attaching screws as you used in the reassembly of the throttle valves, page 81.

Reattach the horizontal link between the two choke shafts using an R-clip. Slip the white plastic seal into its housing and cover it with the white plastic plug. Slip the vertical operating link (heavy wire) through the seal and use an R-clip to reattach the link to the actuating lever. The ends of the vertical lever point inward, toward the carburetor body.

Begin reassembling choke housing internals by slipping the weighted white plastic fast-idle cam onto the tubular bushing. Slip the thrust washer on the bushing and then press the assembly into the choke housing.

Slip a Teflon sleeve onto the choke shaft, then the white plastic choke lever (with its captured return spring). Place the choke shaft into the tubular bushing and reattach the long return spring.

Fit the actuating lever to the backside of the choke housing, then install a washer and locknut. Tightening the locknut puts considerable pressure on the assembly and will easily bend the two long arms at the other end of the shaft if they are not held with pliers.

Refit the pull-off diaphragm assembly by placing the piston in its bore and snapping the retaining collar in place at the same time. The longer of the two limit-arms must rest against the back side of the piston. Test the assembly by pulling out slightly on the diaphragm. The piston should press against the limit-arm and rotate the fast-idle cam/weight assembly slightly as the diaphragm is pulled out. Refit the spring and diaphragm cover. Tighten the three shoulder screws incrementally so the diaphragm is not wrinkled.

Check the O-ring that seals the vacuum passage between the float bowl cover and the choke assembly. Replace it. Then, position the three mounting screws (with their washers and lock washers) that will hold the choke assembly to the float bowl cover. Drop them in place with needle-nose pliers. Next, attach the choke lever to the vertical link from the choke valves. Finally, position the choke assembly on the float bowl cover and secure it with the three screws.

Fit the compression ring and the three screws that hold the bimetallic coil to the choke housing. Match up the index marks on the choke and coil housings, then tighten the screws holding them together.

Float Bowl Cover—Place the float bowl gasket on the cover. Refit the float assembly, with the needle valve, to the float bowl cover. Press the fulcrum pin into its bosses.

Always check the float level whenever the carburetor is disassembled for repair. Measure the distance from the top of the float to the bottom surface of the float bowl cover, including gasket, when the

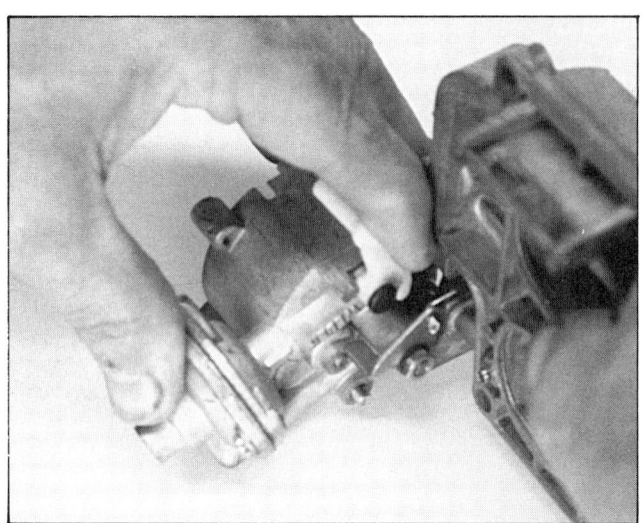
Staked vertical link must be engaged with choke lever during reassembly. Rotate choke assembly to slip link into lever, then position assembly so it can be attached to float-bowl cover.

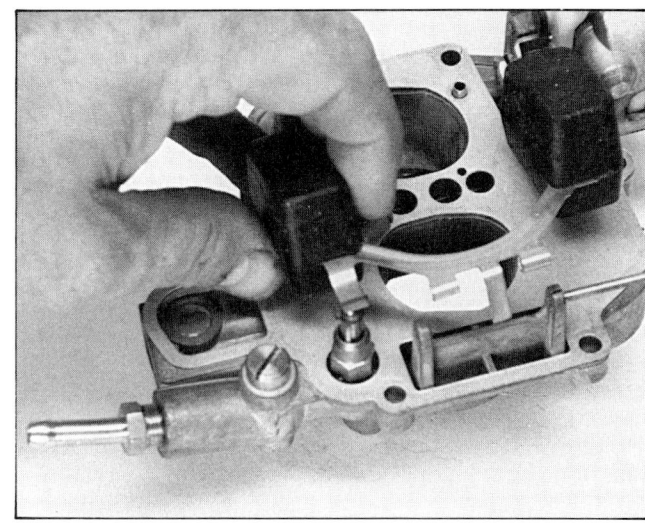
Reassemble floats to float-bowl cover. Float needle valve must hang from float assembly and drop into seat as floats are lowered in place. Float-level settings are done with gasket in place and tab just touching ball on needle valve: closed should be 7mm; open, 16mm. Measure from gasket surface to point on float *nearest* gasket.

float arm is just touching, but not depressing the ball in the float assembly.

This measurement is most easily made by holding the cover vertically with one hand so the float dangles straight down. Angle the cover slightly so the float arm just touches the float needle assembly. Then measure the distance between the float and gasket surface. The correct dimension is 7mm.

Correct the float level when necessary by gently bending the float-arm tab that presses against the float needle.

Refit the float bowl cover to the main carburetor body and reattach it with the six screws. Move the fast idle screw aside when setting the cover on the body, then insert it into the choke-housing cavity after the cover is seated. Check the choke linkage to be sure it works freely.

DUAL-SIDEDRAFT WEBER: DCOE

If you're thinking of Weber performance carburetion, you're probably thinking of the DCOE. As much as any other carburetor, the DCOE typifies the no-holds-barred, out-and-out performance carburetor.

DFT INITIAL SETUP & TUNING
Cold Engine:
- If float level has not been set, do it now.
- Check for fuel leaks. If using an electric fuel pump, turn on ignition so it operates and check all fuel fittings for leaks. If mechanical pump is used, disconnect high-tension (large) wire from coil, crank engine for several seconds and then check for fuel leaks. Reconnect coil wire.
- Verify fuel cut-off solenoid is operating by switching ignition off and on, and listening at carburetor for "click" of solenoid.
- Remove air cleaner, if necessary. Open throttle all the way and then close it. The choke butterfly valve should close completely on cold engine. Move choke butterfly valve linkage to verify there is no binding. If there is any hint of binding, carefully free linkages so operation of choke is correct. Be careful, if you bend a link, not to upset choke setting.
- Hold choke valve open and locate accelerator-pump outlet in primary venturi. Open throttle, then close it. Fuel should flow from accelerator-pump outlet. If it does not, correct problem.
- Have a helper fully depress accelerator pedal. Verify throttle plates fully open. If not, readjust linkage so they open completely.

Idle-mixture richness is not adjustable on the DFT without removing a tamper-proof plug. If an HC/CO meter is available, check carburetor's adjustment. There is little reason to adjust idle mixture of a new DFT. When engine is fully warm, adjust idle speed to 850 rpm, or as specified in the owner's manual or underhood label.
- Disconnect tachometer.

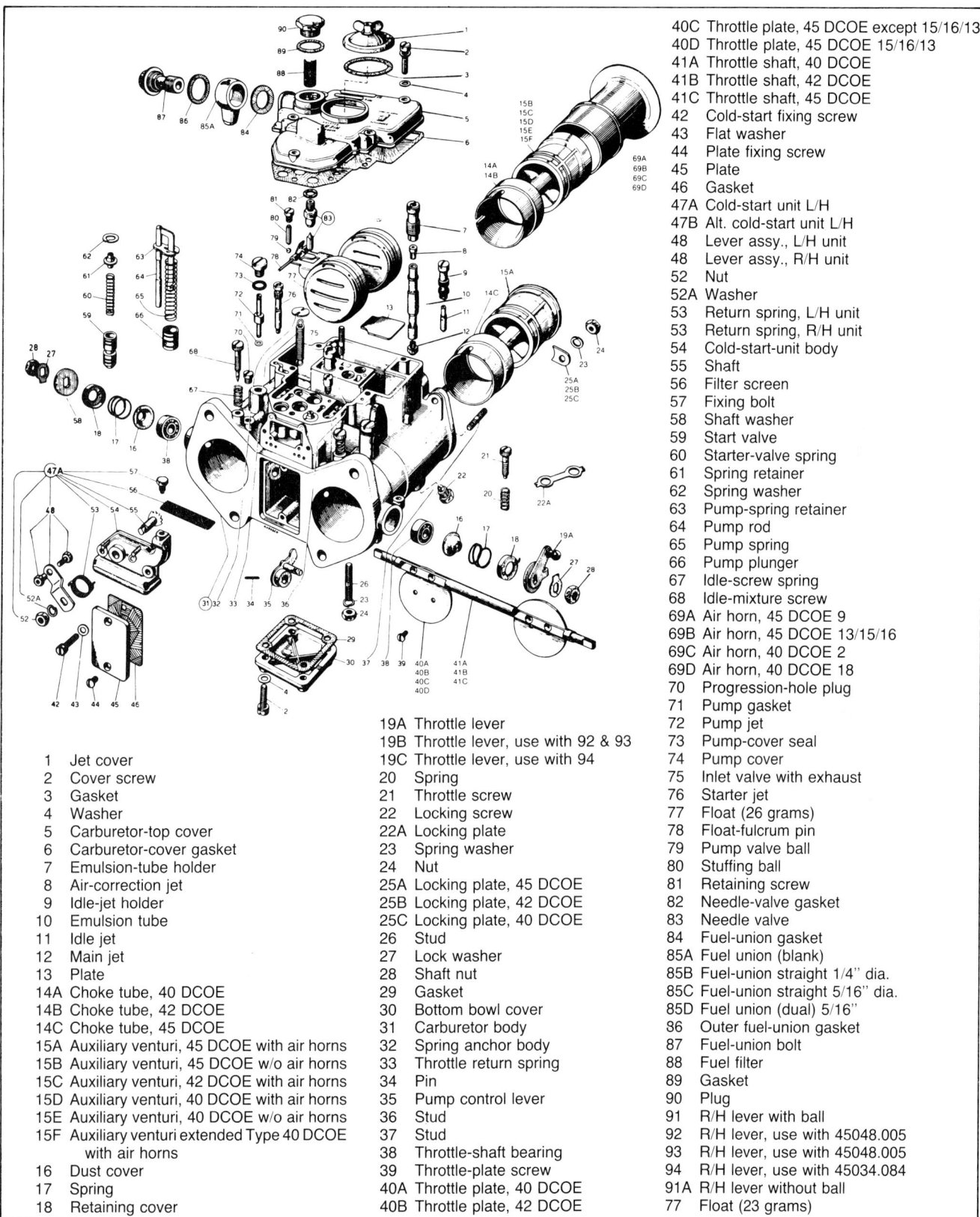

40C	Throttle plate, 45 DCOE except 15/16/13
40D	Throttle plate, 45 DCOE 15/16/13
41A	Throttle shaft, 40 DCOE
41B	Throttle shaft, 42 DCOE
41C	Throttle shaft, 45 DCOE
42	Cold-start fixing screw
43	Flat washer
44	Plate fixing screw
45	Plate
46	Gasket
47A	Cold-start unit L/H
47B	Alt. cold-start unit L/H
48	Lever assy., L/H unit
48	Lever assy., R/H unit
52	Nut
52A	Washer
53	Return spring, L/H unit
53	Return spring, R/H unit
54	Cold-start-unit body
55	Shaft
56	Filter screen
57	Fixing bolt
58	Shaft washer
59	Start valve
60	Starter-valve spring
61	Spring retainer
62	Spring washer
63	Pump-spring retainer
64	Pump rod
65	Pump spring
66	Pump plunger
67	Idle-screw spring
68	Idle-mixture screw
69A	Air horn, 45 DCOE 9
69B	Air horn, 45 DCOE 13/15/16
69C	Air horn, 40 DCOE 2
69D	Air horn, 40 DCOE 18
70	Progression-hole plug
71	Pump gasket
72	Pump jet
73	Pump-cover seal
74	Pump cover
75	Inlet valve with exhaust
76	Starter jet
77	Float (26 grams)
78	Float-fulcrum pin
79	Pump valve ball
80	Stuffing ball
81	Retaining screw
82	Needle-valve gasket
83	Needle valve
84	Fuel-union gasket
85A	Fuel union (blank)
85B	Fuel-union straight 1/4" dia.
85C	Fuel-union straight 5/16" dia.
85D	Fuel union (dual) 5/16"
86	Outer fuel-union gasket
87	Fuel-union bolt
88	Fuel filter
89	Gasket
90	Plug
91	R/H lever with ball
92	R/H lever, use with 45048.005
93	R/H lever, use with 45048.005
94	R/H lever, use with 45034.084
91A	R/H lever without ball
77	Float (23 grams)

1	Jet cover
2	Cover screw
3	Gasket
4	Washer
5	Carburetor-top cover
6	Carburetor-cover gasket
7	Emulsion-tube holder
8	Air-correction jet
9	Idle-jet holder
10	Emulsion tube
11	Idle jet
12	Main jet
13	Plate
14A	Choke tube, 40 DCOE
14B	Choke tube, 42 DCOE
14C	Choke tube, 45 DCOE
15A	Auxiliary venturi, 45 DCOE with air horns
15B	Auxiliary venturi, 45 DCOE w/o air horns
15C	Auxiliary venturi, 42 DCOE with air horns
15D	Auxiliary venturi, 40 DCOE with air horns
15E	Auxiliary venturi, 40 DCOE w/o air horns
15F	Auxiliary venturi extended Type 40 DCOE with air horns
16	Dust cover
17	Spring
18	Retaining cover
19A	Throttle lever
19B	Throttle lever, use with 92 & 93
19C	Throttle lever, use with 94
20	Spring
21	Throttle screw
22	Locking screw
22A	Locking plate
23	Spring washer
24	Nut
25A	Locking plate, 45 DCOE
25B	Locking plate, 42 DCOE
25C	Locking plate, 40 DCOE
26	Stud
27	Lock washer
28	Shaft nut
29	Gasket
30	Bottom bowl cover
31	Carburetor body
32	Spring anchor body
33	Throttle return spring
34	Pin
35	Pump control lever
36	Stud
37	Stud
38	Throttle-shaft bearing
39	Throttle-plate screw
40A	Throttle plate, 40 DCOE
40B	Throttle plate, 42 DCOE

40/42/45 DCOE, drawing courtesy Redline, Inc.

Alfa Romeo "GTA sovraalimentazione" used two turbochargers driven by engine oil pressure—yes, oil pressure!—to get 220 HP from 1600cc and two DCOE carbs. Twin-plug head helped. Photo courtesy Alfa Romeo Archives.

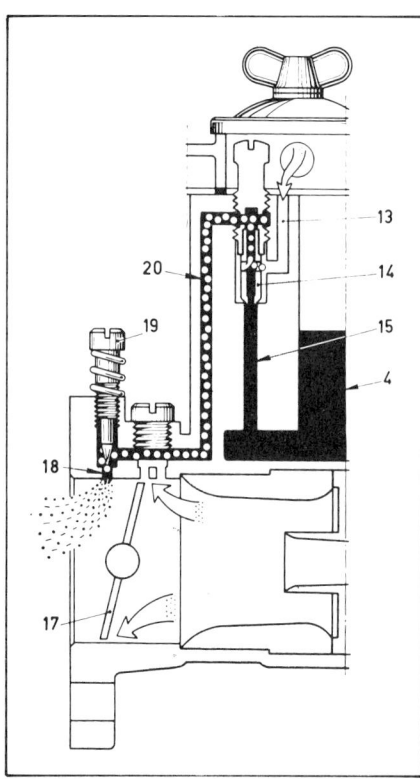

DCOE idle circuit is straightforward; receives its fuel directly from float bowl (4), not through main fuel jet. Drawing courtesy Weber.

Part of this reputation comes from the fact that, as a horizontal carburetor, the DCOE is distinctively different from the exclusively downdraft American carburetor. The reason for the difference is simple. Most American cars have V-type engines, which immediately suggest some kind of carb sitting vertically between the two cylinder banks. Many European engines, on the other hand, are typically in-line fours or sixes, and these engines just as logically accept a sidedraft carburetor. Weber DCOE carburetors have been standard equipment on Alfa Romeo and Maserati in-line engines, as well as virtually every Italian special from Abarth to OSCA. Even the high-performance version of the British Aston Martin sports DCOE carburetors.

The 40 DCOE carburetor is one of that family of sidedraft Weber carburetors with two horizontal venturis and one throttle shaft. The DCO terminology comes from *Doppio Corpo Orrizontale* in Italian, which means, literally *double body—horizontal*. The 40 designation refers to the diameter of the throttle bore, and venturis smaller than 40mm in diameter are fitted. Frequently, numbers follow the DCOE to specify a series or special application.

The main casting of the DCOE includes the two venturis, float bowl, and the jets and their drilled passages. This is a simple carburetor to work on, because most of its removable parts are accessible with the float bowl cover removed.

The ability to field strip a DCOE without removing it from the vehicle has made it a favorite of enthusiasts and manufacturers. It is essentially two carburetors in a single casting because the

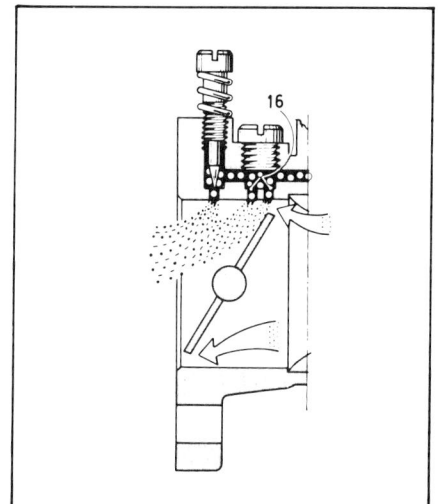

Progression holes supply critical off-idle mixture for optimum driveability. Note that air escaping down bottom side of throttle plate has no fuel mixed with it. Drawing courtesy Weber.

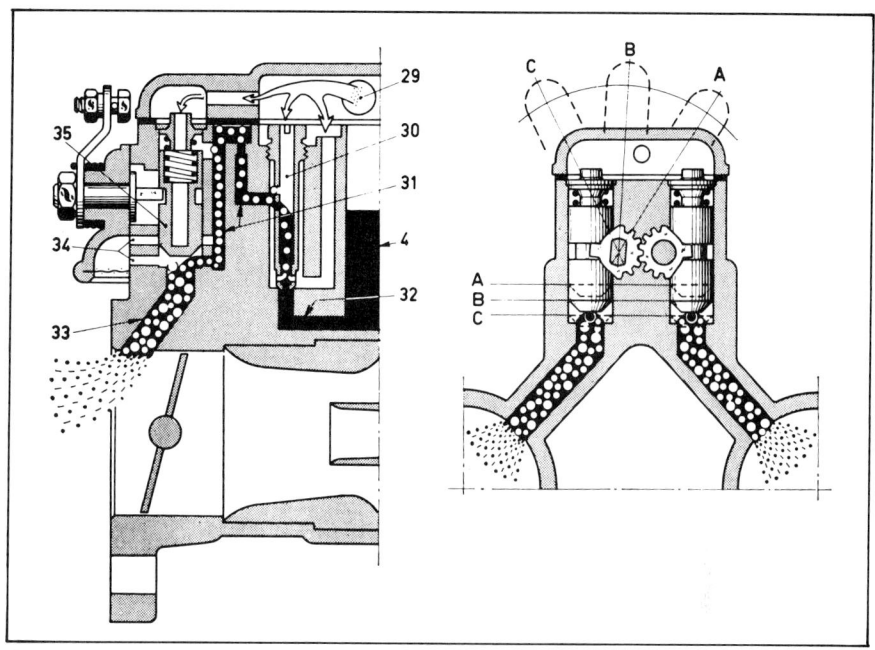

Starting circuit provides dose of extra-rich mixture through its own large passage downwind of throttle plate. Separate choke circuit on DCOE and other Webers is used that eliminates inlet restriction caused by traditional butterfly-valve choke.

two venturis and their associated jets work independently. Only the throttle plates work in unison. And the single throttle-shaft design ensures they are always synchronized.

The DCOE is unique in that its cold-start mechanism is somewhat like a separate internal carburetor. It, plus an accelerator pump, enrich the fuel mixture as needed. There is no power enrichment valve or emission control device fitted to the DCOE, and it is regarded as strictly a performance carburetor. Consequently, most of its applications are for off-road and racing.

Rejetting this carburetor is simple because all jets are accessible from the top. The main and idle jet assemblies are carried under the central *hat*, which is held on by a thumbscrew. Once removed, the hat reveals the jet assemblies, which can be removed with a broad-bladed screwdriver. The assemblies carry both the main jet, at the very bottom, the emulsion tube, which makes up the major part of the shaft, and the air-correction jet, which is underneath the holder you un-

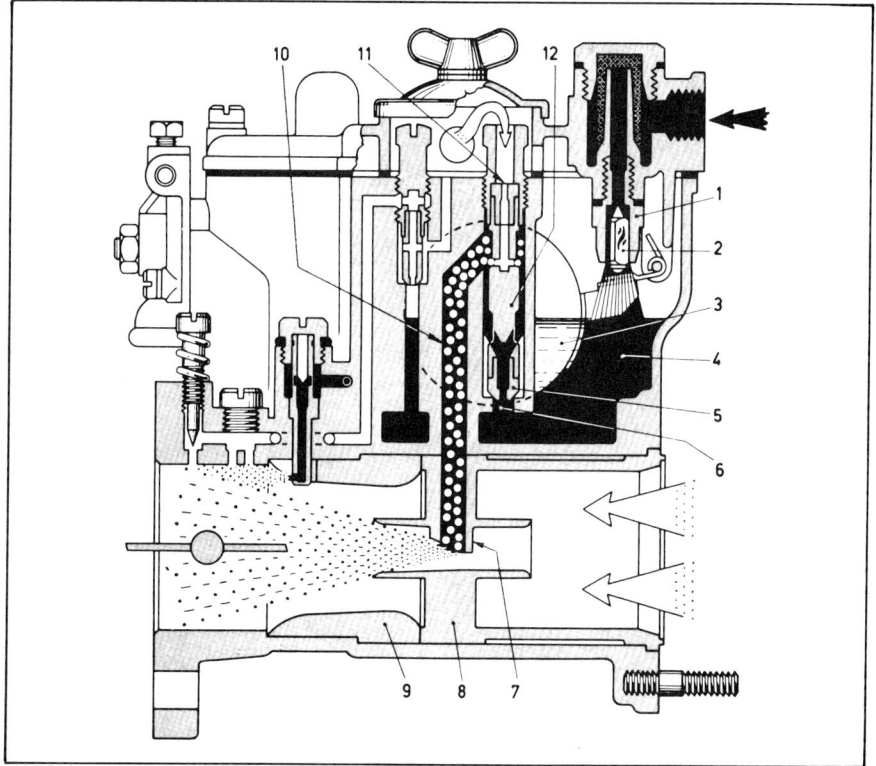

Main circuit is also straightforward and relatively short: good response to throttle is major benefit. Drawing courtesy Weber.

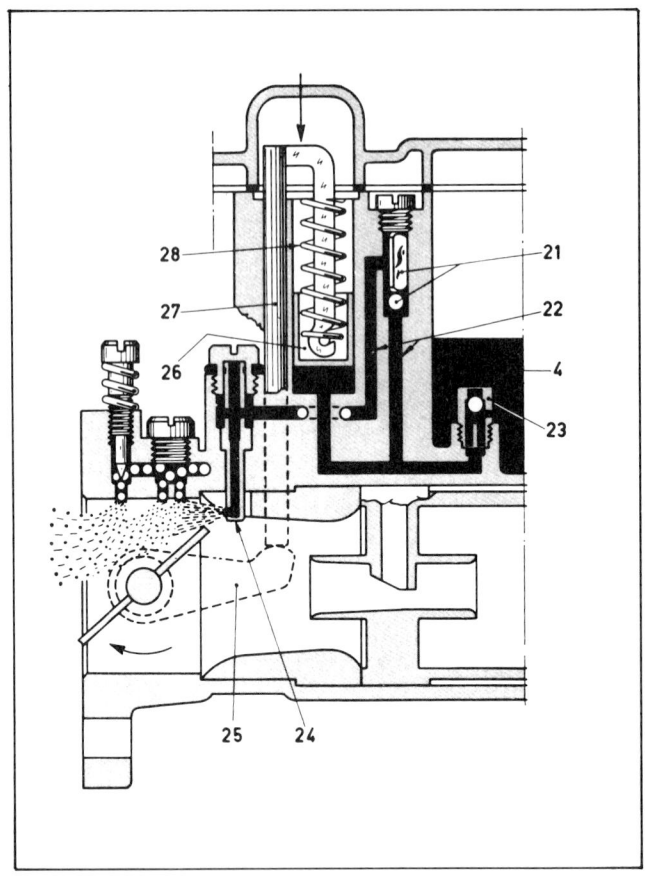

Accelerator-pump circuit includes piston, two check balls. Circuit is unique in that fuel is also pulled through it during full-throttle operation.

screw when removing the assembly. All the pieces pull apart: four sections for the main jet and two sections for the idle jet.

With the DCOE, you can easily make up sets of jet assemblies to give predetermined air/fuel profiles, and then use them to quickly fine-tune the engine for whatever fuel delivery characteristics are optimum.

It's not practical to disassemble this carburetor completely just to fix one part. You can do more accidental harm disassembling a DCOE than could ever result from normal wear on an engine. A little extra time spent troubleshooting can save the grief of breaking a perfectly good part during disassembly.

The 40 DCOE throttle shaft is carried in ball bearings to eliminate wear of the carburetor body. Replacing these bearings means removing the throttle valves, the most challenging repair operation of all.

If the bearings are not worn, refurbishing a DCOE is straightforward enough that it can be performed without removing the carburetor from the car. If they are worn, consider replacing the complete unit or sending it to a Weber specialist for repair. The cost could be less because throttle shaft renewal is an exacting exercise.

Choke assembly removal requires dismounting the carburetor in some installations to obtain clearance to remove the attaching screws.

The disassembly and reassembly procedure is divided into sections. It covers the complete renewal of the carburetor when it's off the car. It includes:
- Removal of float bowl cover.
- Removal of components accessible when float bowl cover is removed, including the accelerator pump.
- Disassembly of starting assembly on outside of carburetor.
- Removal of throttle shaft.

The procedure only suggests the order in which the DCOE carburetor can be disassembled. After removing the float bowl cover, any major component can be disassembled without disassembling any other. Your disassembly process should depend on the repair needs of the carburetor, and should avoid unnecessary disassembly.

Fuel-filter cover, filter are easily removed for frequent in-car cleaning. Install non-restrictive in-line fuel filter for extra insurance.

Jet inspection cover is famous Weber "Mickey Mouse hat." Main- and idle-jet assemblies are easily accessible beneath.

Five screws hold float-bowl cover to DCOE. Gasket removes with cover.

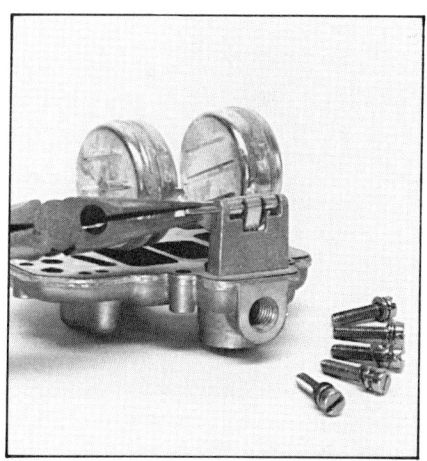
Float assembly is removed by pulling pin with needle-nose pliers. Needle valve is held on float assembly by small spring.

DISASSEMBLY

See the sidebar, page 45, for information on special Weber tools and their application. If you're going to do a complete overhaul, buy an overhaul kit. If you're doing anything less than a complete overhaul, try to save most of the old gaskets and diaphragms necessary to successfully reassemble the DCOE. If you're trying to save a gasket or diaphragm, work very slowly and carefully to separate it from its mating surface.

Begin the overhaul by cleaning the outside of the carburetor as completely as possible. Scrape off baked-on dirt with a knife or screwdriver, but be careful not to gouge out pieces of the carburetor in an attempt to clean it. Use an aerosol can of carburetor cleaner to remove varnish and any remaining dirt. Follow the instructions and precautions on the cleaner's can. Use goggles and avoid breathing the spray.

It's the mark of a professional to make sketches and notes during disassembly. Operations such as dismantling a throttle shaft create lots of parts that must be reinstalled in their correct order. Any diagrams, no matter how crude, are helpful when facing a handful of levers, washers and springs.

Float Bowl Cover—Remove the fuel filter, clean it thoroughly and set it aside for reassembly. Remove the jet inspection cover. Remove five float cover screws, washers and lock washers, then lift off the float cover, being careful not to tear the gasket.

If you can't remove the cover with a slight pull, turn the carburetor over and use the screwdriver handle to tap the cover and loosen it from the carburetor body. Hit the cover lightly. The gasket usually stays attached to the float bowl cover, though it may be attached to the cover at one point and the main body at another.

To save the gasket, slide a knife along the carburetor body side, separating the gasket from the mating surface as slowly and carefully as possible. The gasket remains with the float bowl cover because it's trapped by the floats and can only be removed completely after the floats have been disassembled from the cover.

The gasket may be very brittle from age and heat; it's not always possible to remove one whole. If you do tear the gasket, replace it.

Turn the cover over, plug the filter opening with your finger, and blow into the fuel inlet. With only the weight of the float assembly pressing on the needle valve, you should not be able to blow air past the seat. If you can, and there's no dirt trapped between them, the needle and seat will have to be replaced.

Float—Use needle-nose pliers to pull the pin holding the float assembly to the cover. Pull the pin out through the solid boss, not through the split boss. Lift off the float assembly and extract the needle valve.

Unscrew the needle-valve seat. This

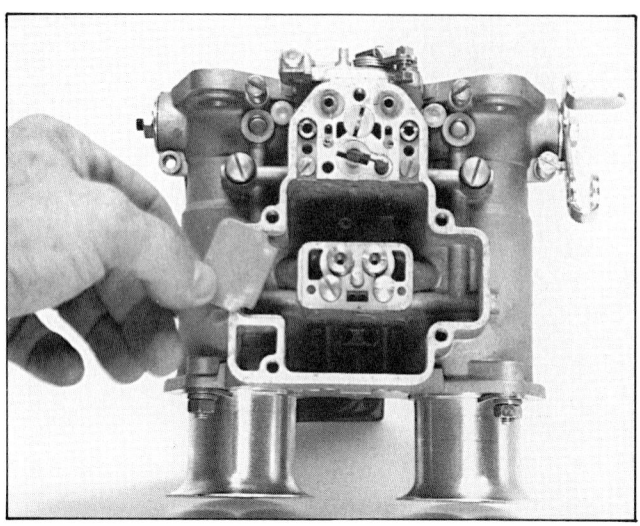
Float-bowl baffle plate lifts from its seat. Baffle reduces danger of fuel sloshing up into air-correction jets.

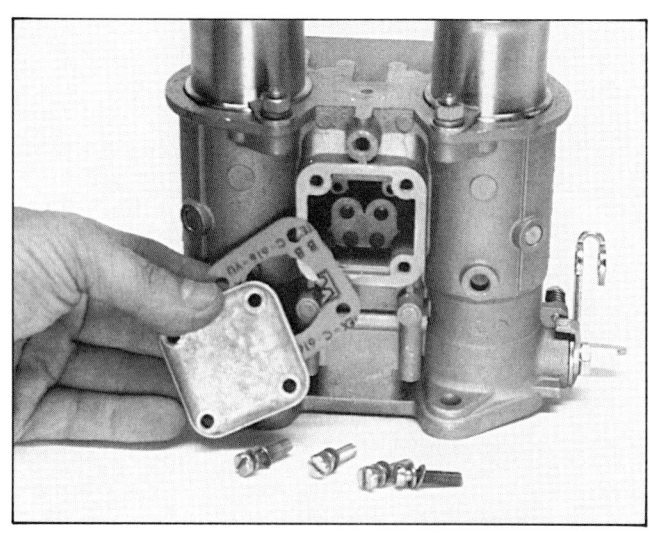
Float-bowl bottom unscrews for thorough cleaning. Gasket should always be replaced to prevent fuel from leaking.

Air horns add beauty, some degree of tuning to DCOE. They retain venturis, so engine should never be run without them.

Main-jet assembly should be unscrewed carefully with correct-size screwdriver. Blade should just fit slot. Main jet is bottom-most element of pull-apart assembly.

step is necessary only if you're replacing the needle and seat as a set. The gasket under the valve seat is shiny, and can look like it's part of the aluminum casting. Double-check to be sure you've removed the gasket. If you don't and put another one on top of it during reassembly, the float level will be changed.

Remove the float cover gasket. Shake the floats and listen to them for sloshing fuel. If you hear any, the floats leaked and must be replaced. Or, weigh the floats to determine if they have absorbed fuel. DCOE floats weigh 26 grams, so if they weigh more, replace them.

Remove the carburetor-bowl baffle plate.

Remove the four float bowl bottom plate screws, bottom plate and gasket if necessary.

Air Horns—Remove the air horns, if fitted, and you plan to remove the venturis. Because the air horns hold the venturis in place, always replace them before restarting the engine.

Jets—Two idle and two main jet assemblies are in the casting protruding from the float bowl. Unscrew each one carefully with a screwdriver that snugly fits the slots.

Idle-jet circuits on sectioned DCOE: Two passages for idle-circuit air correction are cut at approximately 45° angle in center of photo. Idle circuit runs on either side of jets, make 90° turn outboard, then another 90° turn to run parallel to venturi.

Idle-mixture-screw removal: Note number of turns to seat, then unscrew. When installing, set to number of turns out from seat for intial adjustment.

Remove U-shaped bronze retainer just before lifting accelerator pump from body.

The idle jets are pressed into the bottom of the two smaller and shorter assemblies. The jets can be separated from their holders simply by pulling them from the threaded upper section. Blow through the jets to clean them. Never use wire or a drill to clean the jets.

The larger two assemblies can be pulled apart into four pieces: 1) the threaded emulsion tube holder, 2) air-correction jet, 3) emulsion tube and 4) main jet. The assemblies may be hard to pull apart. If you have to use pliers as a last resort, put cloth tape on the jaws to avoid scratching the jets.

Idle Mixture Screw—Record the number of turns required to gently seat each idle mixture screw. Then unscrew it from the carburetor body. Remove each screw with its spring, washer and, if it's on the screw, remove the O-ring. When replacing the screw, use the number of turns from the seated position to make the initial adjustment. Remove the idle speed adjustment screw if necessary.

Accelerator Pump—Remove the accelerator pump assembly from the carburetor body by prying up the U-shaped bronze piece (retaining plate) with a small screwdriver.

One check ball for accelerator-pump circuit is removed after retaining screw is off.

Accelerator-pump-circuit discharge valve is at bottom of float bowl.

Accelerator-pump jet is held in place by cap screw. Jet is centered in its bore by large middle section. One side of middle section has a flat to position jet outlet so it points downwind.

Sectioned DCOE: Starter-valve housing is at lower left. Rightmost sectioned "tower" carries main- and idle-jet assemblies, with main jet wells exposed.

If necessary, dismantle the accelerator pump assembly by removing the piston from the operating rod. Press and twist the piston. It will come off the hook at the end of the rod. Then remove the spring and plate.

Remove the check-valve retaining screw, spacer and ball at the top of the casting. Then unscrew the discharge valve at the bottom of the float bowl.

Both the check valve and discharge valve are part of the accelerator pump circuit. Check the discharge valve for one-way operation by blowing through it. It should pass air only one way.

Unscrew the cap screw for the accelerator-pump jet, then remove it, along with the washer and jet. A rubber gasket is captive with the jet.

Starter Circuit—Remove the two starter circuit jets. Remove the two screws holding the starter circuit mechanical control assembly to the carburetor body and remove it. The control assembly includes two geared levers that lift the starter valves (pistons) in the main body of the carburetor.

If necessary, dismantle the starter circuit mechanical control assembly by unscrewing the nut from the shaft, then

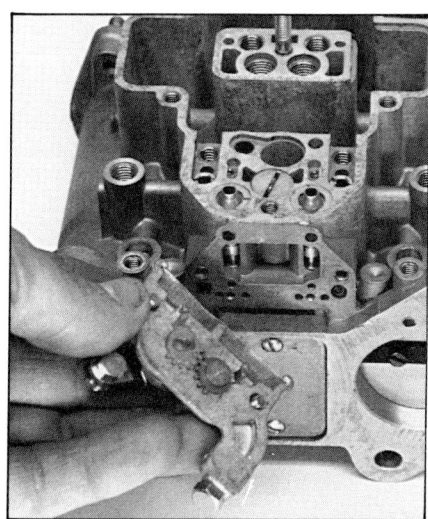
Mechanical control assembly for starter circuit features two gear-meshed tabs. Tabs lift piston in circuit to pass air/fuel mixture.

Starter-circuit jet is somewhat recessed. Top of jet is easily scarred unless correct-size screwdriver is used to extract it.

Release spring washers that hold starter valves in place, under spring pressure. Be careful not to bend washers.

remove the outside lever, spring and the two geared levers. Be careful not to damage the flat, fine-screen air filter. It can be pulled out for cleaning. Leave the assembly intact if it is working. A cleaning is usually all that's required.

Mark the starter valves with pencil so you can refit them in their original bores. Remove the starter circuit valves by slowly and carefully prying up the spring washers with a small screwdriver; then pull up each assembly, which includes the retainer/guide, spring and valve.

If there is one, remove the blanking screw for the progression holes. Some carburetors have a brass plug instead of a screw, which can't be removed.

Remove the two screws holding the rear plate, then remove it with its gasket.

Throttle Valves—Don't remove the throttle valves and shaft unless **absolutely necessary**. The chances of making errors and ruining parts are high. So, what's absolutely necessary? If a valve is binding and can't be freed, or its action improved with cleaner, the shaft is bent or worn. Consider, too, that you must also remove the auxiliary venturi and venturi to reinstall the throttle valves.

Section at throttle plate of DCOE shows idle, progression circuits. Throttle plate is completely closed. Just above it are four progression holes, two of them revealed here. Accelerator-pump-outlet passage has been drilled out farther upwind of progression holes.

Removal of throttle-shaft screws is not recommended; should be performed only if absolutely necessary. Operation is delicate, prone to stripping threads in throttle shaft.

Return spring is strong: Needle-nose pliers must be used to release it from lever.

Split pin must be removed to free throttle return-spring lever from throttle shaft.

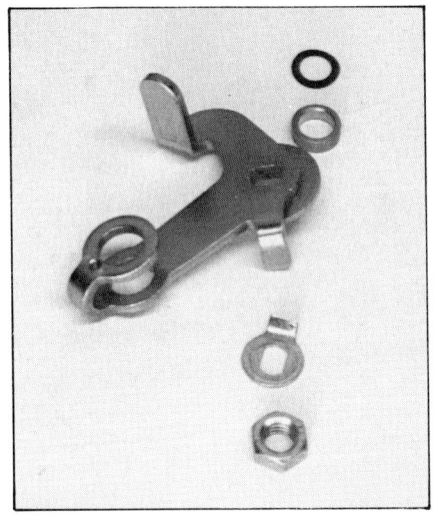

Throttle-shaft end pieces removed in order. Orientation of lever doesn't match actual position on shaft.

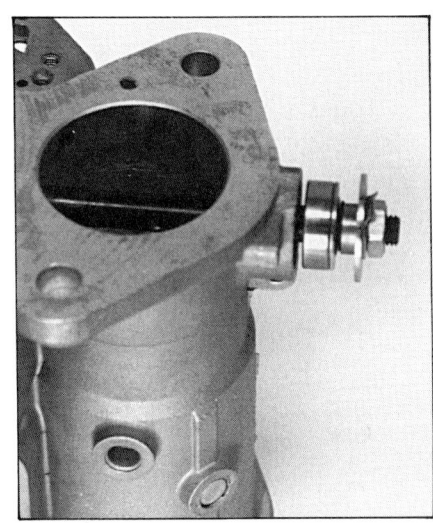

Throttle-shaft bearing tapped free of its boss: Shoulder on shaft seats against bearing inner race.

A worn throttle shaft will produce an erratic idle speed or an inconsistent return to a stable idle. If faced with these conditions, either replace the complete unit or send it to a Weber specialist to do the repair. I've included the removal and assembly procedures for those with no fear, a surgeon's dexterity, and plenty of time.

The first step is to remove the four screws that hold the throttle valves to the shaft. The ends of the screws are staked to keep them from vibrating loose. Only slight torque can be applied to the screws with a screwdriver to avoid bending the throttle shaft. If slight torque won't remove the screws, grind their ends so they can be removed without stripping the threads. The shaft is hard, and so even if you do strip the screw, there should be enough thread left in the shaft to chase with a 4 × 0.7mm tap so that new screws can be used.

Mark both throttle valves with a pencil so they can be replaced exactly as they were removed; then remove the valves.

Throttle Shaft—Closely inspect the throttle shaft to see that no burrs were caused when the screws were removed. If there are any burrs, they must be removed now with a file. This is so the shaft bearings can be removed easily without cutting the shaft seals.

Remove the throttle-shaft return spring along with its anchoring plate from the cavity between the venturi bores. The spring is under significant tension. Lift up on it with needle-nose pliers at its top end and slip the anchor plate free. Remove the spring from the shaft lever using needle-nose pliers.

The throttle shaft lever is pinned. Drive out the *split pin* with a punch.

Bend back the lock tabs at the ends of the throttle shaft and remove the 11mm nuts, tab washer, lever, thick washer, thin washer, spring covers, springs and dust covers.

Tap the throttle shaft free. One ball bearing will come with it. Remove the throttle-shaft return lever from the center cavity.

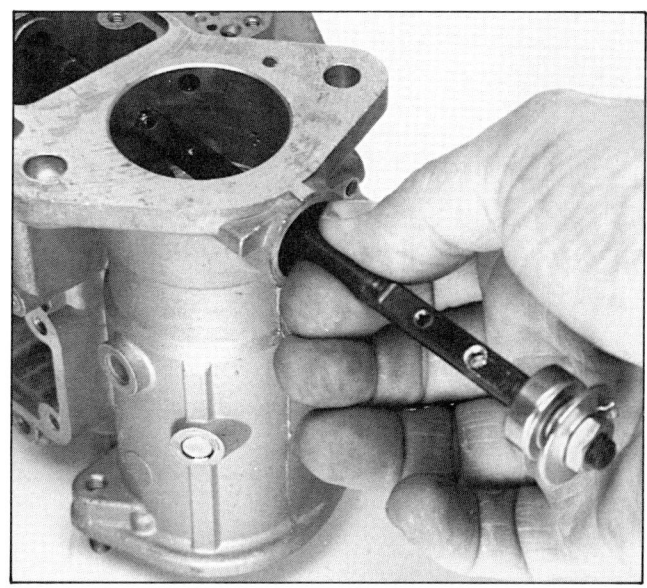

Remove second throttle-shaft bearing. Shaft should pull out easily once bearing is free.

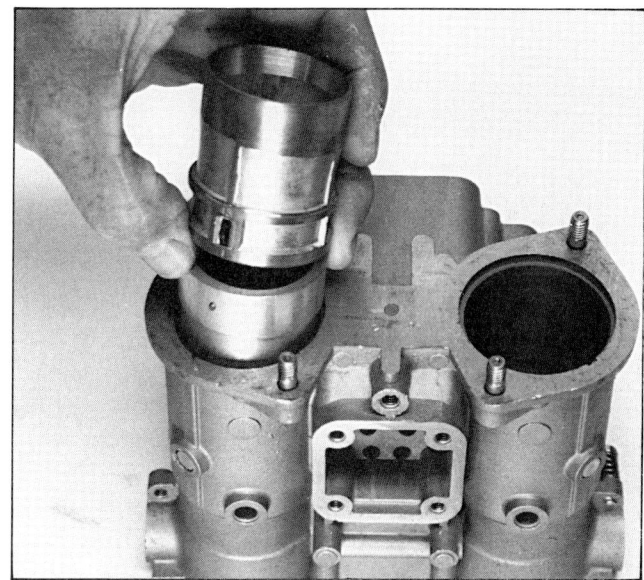

Venturi removal: Auxiliary venturi and main venturi stack into throttle bore. Auxiliary Venturi is held in position with small spring that fits in guide. Main venturi may be retained by lock screw in some applications.

Remove the bearing from the throttle shaft, refit the throttle shaft to the carb body and use it as a driver to remove the other ball bearing. Be careful not to damage the threads on the throttle shaft.

Venturis—Remove the venturi retaining screws and locknuts, if there are any (found in 45 DCOEs and some 40 DCOEs), and then remove the auxiliary venturis and venturis by lifting them out of their bores. Note that they go in only one way because they have locating pins (venturis) or clips (auxiliary venturis).

If you have removed the throttle valves, it's *absolutely necessary* to remove the venturis to permit correct reinstallation of the plates. If the venturis won't lift out easily, tap them out using a wood dowel from underneath. Weber has an expensive tool for removing stuck venturis if they can't be tapped out using the dowel.

INSPECTION

Cleaning—See the Cleaning section, page 51, of the IMPE part of this chapter for tips on cleaning the carburetor.

Inspect all of the drilled passages in a carburetor as you clean them with the spray or an ear syringe full of solvent. It's essential that you wear goggles when you do this. Work in a well-ventilated area away from open flame or pilot lights.

Aerosol cleaners come with a plastic tube that inserts into the spray nozzle. Use the tube to probe the passages while spraying cleaner. It's unusual for passages to be blocked. Usually, dirt blocks the jets before the passages. A more common problem is the loss of plugs used to block the ends of drilled passages.

The relative low cost of a Weber carburetor, including the DCOE series, is one reason not to do extensive or heroic repair procedures. Of course, some DCO Webers are old and irreplaceable, but that fact is an argument for sending extensive repair operations to a Weber specialist with the skills and tools.

Float Bowl Cover—The gasket sealing the float bowl cover to the carburetor main casting should show that the sealing surfaces of the two castings are parallel. Check the float bowl cover for broken flanges, cracks or other physical damage. Check the threads for the fuel filter plug and needle-valve seat.

Fuel Filter & Floats—Check the fuel filter for any damage to the screen. Blow it clean with compressed air to remove any dirt. Install a low-restriction in-line fuel filter as extra insurance to keep dirt from reaching the internals.

The floats shouldn't show any evidence of leaking. Shake the assembly near your ear and listen for gasoline sloshing around inside. If the floats leaked, replace them.

The float needle valve and its seat should show even wear where they seat against each other. If not, replace them.

Main Casting—Blow air through all the passages you can identify, including those supplying fuel to the main jets, idle jets, intermediate circuits, accelerator pump and starting circuit. Never use a wire or drill to probe or clean the jets.

Remove the idle jets from their holders and check each by holding it up to a light. Blow through the jet if it's clogged.

Blow air both directions through the accelerator-pump discharge valve. You should be able to blow only in one direction because it's a one-way valve.

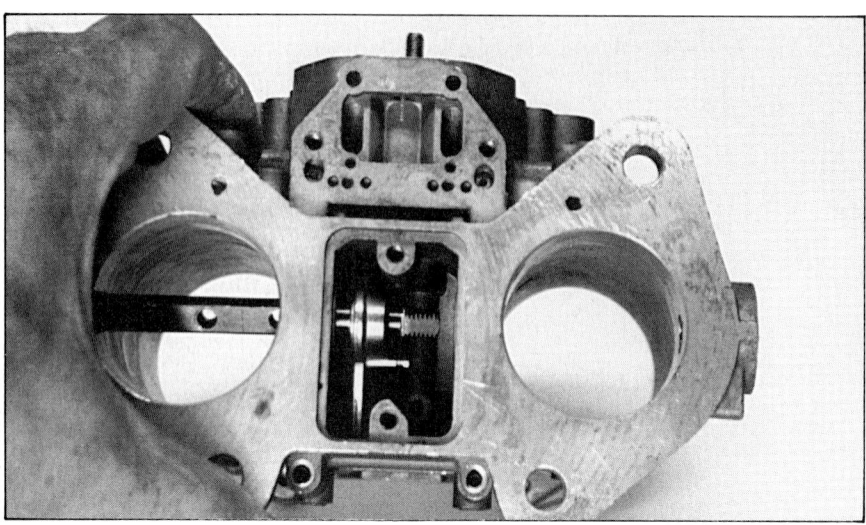

Refitting throttle shaft: Make sure throttle-shaft-return bellcrank is positioned as shown. Installing it backwards is possible, and very frustrating.

Check for any damage to the main casting. Use a straightedge to check the mating surface of the base of the carburetor for warpage.

Auxiliary Venturi—If these have been removed, blow though the passage that delivers fuel to them and remove any burrs or dirt from the venturi.

Throttle Valves & Shaft—The throttle shafts should be straight, with no visible wear where they operate in their ball bearings. The bearings should freely rotate. Check throttle shaft straightness by rolling it on a flat surface. Double-check the threads at the ends of the shaft.

The throttle valves should have smooth edges and be flat. All springs that slide onto the throttle shaft should be strong and not deformed. The nylon dust seals should be replaced if they show any evidence of wear.

REASSEMBLY

The disassembly of a DCOE is, in itself, a lesson in reassembly. After all components have been inspected and damaged parts replaced, begin reassembling any of the major components first. Start with either the float bowl cover or the main casting. Use diagrams and notes you made during disassembly to help. The following text gives the most important points to remember in reassembly.

Throttle Shaft—There are two dangers in reassembling the throttle valves: bending the throttle shaft and not positioning the valves correctly. Overcoming the first requires care; the second, some strategy. The holes in the valves are oversize to permit them to be positioned for a perfect seat in the bores.

Lubricate with light grease, then fit the throttle shaft to the carburetor body, feeding it through the throttle-shaft return bellcrank in the middle cavity.

In the next several operations, keep checking to see that the throttle shaft doesn't bind excessively in the carburetor body. It will move increasingly more freely as its assembly progresses. But it may move with no trace of binding only after the throttle plates have been fitted and their attaching screws tightened. The reason for frequent checks is to avoid a binding throttle shaft by identifying and removing burrs and other problems during assembly.

Lightly lube with grease the ball bearings on each end of the throttle shaft, then fit the bronze-colored washers and nuts to each end. Slowly tighten the nuts to draw the bearings into the casting as far as you can.

Remove the nuts from both ends of the throttle shaft, then fit the dust seals, springs, washers, lever, lock-tab washer and nut to each end of the throttle shaft. Retighten the nuts, forcing the bearings home in their bores. Bend up the lock tabs.

Throttle Valves—Take the appropriate steps to prevent loosening of the throttle valve screws. They can vibrate free and make their way into the engine and cause damage. Use fuel-resistant Loctite on the threads and deform the screw ends. You can deform the screw ends with a chisel only if you support them properly so no force is transmitted to the throttle shaft. Otherwise, you'll bend the shaft and have to replace it.

First, slip the throttle valves into their original positions in the throttle shaft, then very carefully screw in their retaining screws fingertight. (Don't forget the Loctite on their threads.) Check first that the plates seat equally well around the circumference of the bore by holding the carburetor body up to a bright light and looking for light leaks. The light should leak as equally around the bore as you can position the plates. Check again for throttle-shaft binding, then tighten the screws securely. The throttle shaft should now operate freely. If it binds, disassemble the throttle plate and shaft assembly until the cause is found.

Stake the ends of the throttle plate screws. Secure a chisel or punch vertically in the jaws of a vise with the flat end up. Have a helper hold the carburetor on the chisel so that the slotted head of one throttle plate screw rests securely on the chisel with the throttle plates closed. Use a sharp, long chisel to reach down inside the carburetor body. Position it carefully on the end of the same throttle plate screw. Double-check that the force of staking won't be absorbed by the throttle shaft, but rather by the chisel held in the vice. Strike the long chisel with a single, sharp blow to stake the screw.

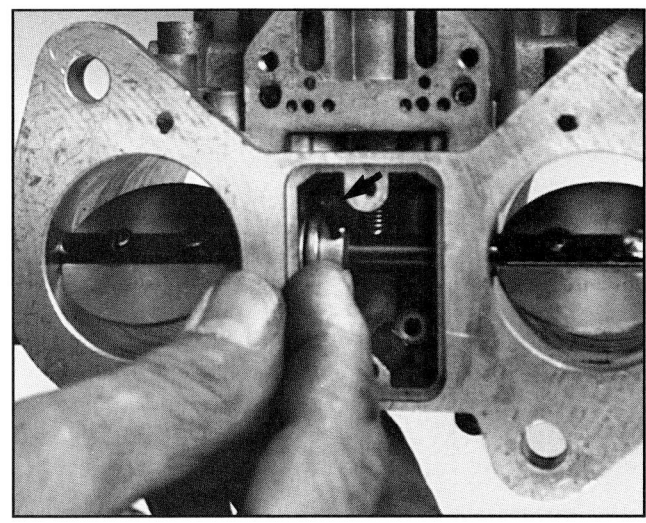

Line up throttle-shaft-return bellcrank with hole in throttle shaft, then fit split pin into bellcrank. Several tries will probably be necessary before split pin can be tapped home.

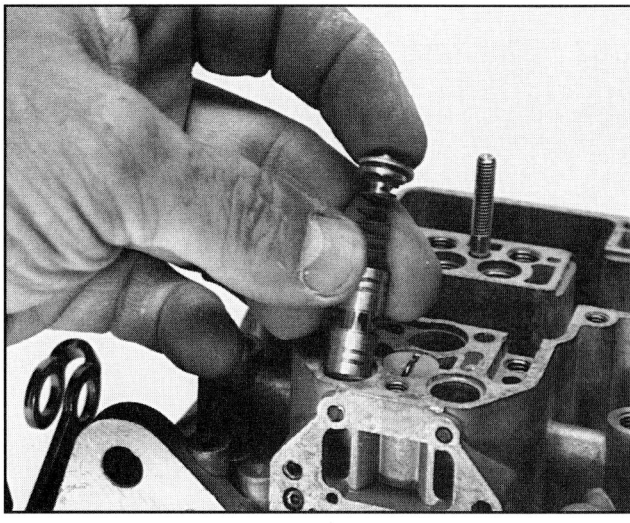

Starter-valve assembly being installed: Parts don't need to be assembled: just drop them into hole in proper order, as shown.

The chisel should create a V-shaped indentation in the end of the screw, expanding the threads at the tip so they are now slightly larger in diameter. Stake the other screws in a similar manner after supporting them against the chisel that's held in the vise.

Check the throttle shaft again for binding. If it binds, you probably bent the throttle shaft during staking and will have to replace it with a new one.

Start the split pin into the throttle-shaft return bellcrank, then slip one end of the return spring onto the bellcrank. Move the bellcrank against the casting so the mating hole in the throttle shaft is visible. Align the split pin with the throttle shaft hole, then move the bellcrank along the throttle shaft so the split pin will enter the hole. Tap it home.

Reach into the carb casting with needle-nose pliers and pull up on the spring. Slip the spring retaining washer beneath the top loop of the spring and let the spring pull it into the recess in the carb body. Refit the gasket and plate over the throttle bellcrank opening.

Venturis—Press the venturis and auxiliary venturis into their bores. They are a one-way fit. Tighten any retaining screws and locknuts used to secure them.

Starter Circuit—Put the starter valves (pistons) in their original bores, then insert the spring, guide and retainer for each. The retainer is a press fit into its recess in the carb body.

Clean the screen on the starter cover, then slip it into the cover. Use a new gasket and fit the starter cover so that the two geared tabs engage the grooves in the starter valves. Tighten the two screws that hold the cover on, then operate the starter to check for binding.

Put the starter jets in their bores near the starter valve assemblies. Refit the pump jets in their bores. The grooved end of the jet goes up. Check that the rubber seal is not damaged, then refit it and the cap screws.

Place the discharge valve into the bottom of the float bowl. Use needle-nose pliers to start the jet into its threads.

Accelerator Pump—Install the accelerator-pump check balls, weights and screws. Refit the accelerator pump piston assembly into the carburetor body, press the retaining plate home in its recess, then operate the throttle shaft to check for binding.

Idle Mixture Screws—Screw in the idle

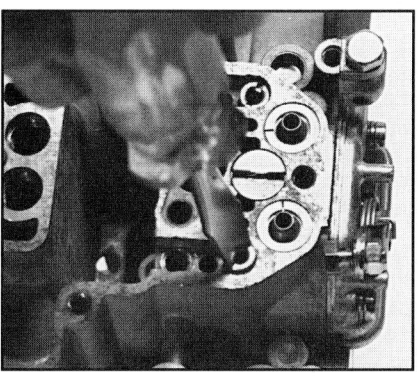

Carefully secure starter jets using a screwdriver that exactly fits their slots. Don't use too much torque and avoid scarring jet.

Unless you have skinny fingers, use pliers to insert accelerator-pump discharge valve into bottom of float bowl.

Accelerator-pump check balls, weights and cap screws in order they install in carburetor body.

Accelerator-pump piston should fit smoothly in bore. Spring must be assembled to rod as shown, with U-shaped retainer in place, before assembly is slipped home.

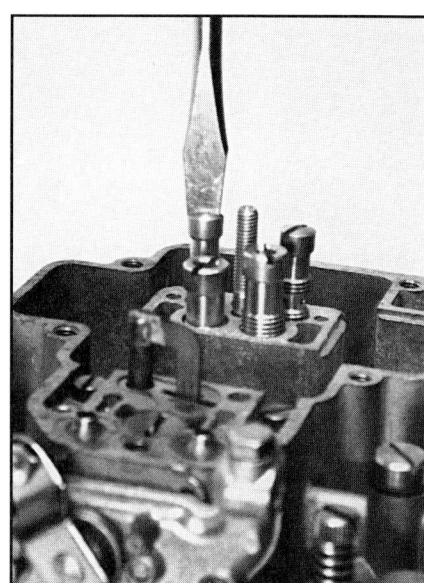
Refit main- and idle-jet assemblies. Only main jet, front left, is fully installed.

adjustment mixture screws, with their springs and O-rings and seat them very lightly. Then, back them out the number of turns they were originally. Or, back them out 3/4-turn for a rough initial adjustment.

If removed, refit the idle speed screw, and its spring. Back it out so it just breaks contact with the throttle lever. Then turn it back in 1/2 turn.

Jets—Hold the idle and main jets up to a light to see that they are clear. If you can't see through the jets, blow them out with air or soak them in carburetor cleaner. Never clean a carburetor jet with a wire or drill. Reassemble the idle jet and main jet assemblies and refit them to the carburetor.

Float Cover Plate—Refit the float cover plate, float-bowl bottom plate gasket and plate.

Float Bowl—If you dismantled the float bowl cover components, refit a new float bowl cover gasket if the old one was damaged. Reinstall the needle seat, needle and floats. Insert the float-fulcrum pin in the solid boss first, then tap it across to the split boss with a suitable punch.

Set Float Level—The carburetor has been apart, so you must check and set float level. Do this by measuring the distance from the float to the bottom surface of the float bowl cover (including gasket) when the float arm is just touching, but not depressing, the float needle assembly. Be accurate.

This measurement is most easily made by holding the cover vertically with one hand so that the float dangles straight down. Move the cover slightly so that the float arm just touches the float needle assembly, then measure the distance between the float and gasket surface using your other hand. For metal floats assembled from two halves, the correct measurement is made to the surface of the float, and not to the raised soldered joint at the circumference of the float.

Correct the float level when necessary by bending the float arm tab that presses against the float needle assembly. On the DCOE carburetor, there is also a limit to the distance the float can drop, adjusted by bending a tab near the fulcrum point. See the accompanying table for typical float setting measurements.

Float Bowl Cover—Reassemble the float bowl cover to the main body. Refit the round jet cover (*hat*) and insert a new fuel filter into the top casting. Use a new gasket under the filter cover bolt. Refit the air horns and secure them with their clamp washers and nuts.

Preliminary Idle Adjustment—After installing the carburetor(s) on the engine, start it, warm the engine to full operating temperature, and adjust the idle as follows. Turn the idle mixture screws in until the engine starts to misfire. If they can go all the way in without causing misfire, recheck the idle speed screw. It is turned in too far.

Gradually turn the mixture screws out until the engine is at its smoothest, fastest idle. After this mixture adjustment, set the final idle speed by turning the idle speed screw until the engine idles at its recommended rpm. Most DCOE-equipped engines will idle best at a CO reading between 3.0 and 4.5%.

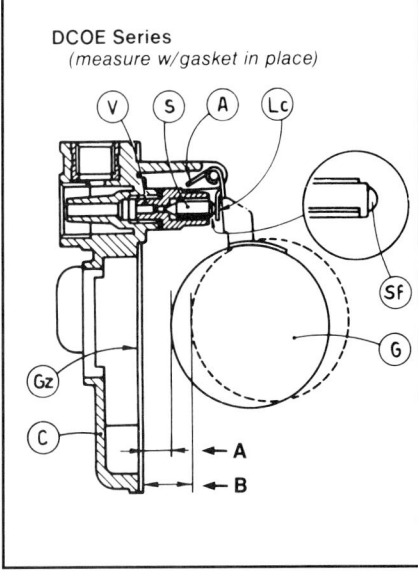

DCOE float-level adjustment is made with gasket in place. Drawing courtesy Redline, Inc.

Float Level Settings

Carburetor	Application	A	B	Stroke
40 DCOE 2	Alfa Romeo Giulietta SV	8.5	15.0	6.5
40 DCOE 2	Alfa Romeo Giulia SS	8.5	15.0	6.5
40 DCOE 2	Lotus Ford Anglia 100 E	8.5	15.0	6.5
40 DCOE 4	Alfa Romeo Giulia Sprint GT	8.5	15.0	6.5
40 DCOE 18	Lotus Elan/Ford Cortina	8.5	15.0	6.5
40 DCOE 20/21	Lamborghini 300 G.T.- 400 GT	8.5	15.0	6.5
40 DCOE 24	Alfa Romeo Giulia Super	8.5	15.0	6.5
40 DCOE 25/26	Renault 8 Gordini	8.5	15.0	6.5
40 DCOE 27	Alfa Romeo Sprint GTV	8.5	15.0	6.5
40 DCOE 28	Alfa Romeo Giulia 1300 GT Jr.	8.5	15.0	6.5
42 DCOE 8	Maserati 3500 GT	5.0	13.5	8.5
45 DCOE 9	Alfa Romeo 2600	7.0	13.5	6.5
45 DCOE 9	Aston Martin DB 4 Vantage GT	5.0	13.5	8.5
45 DCOE 9	Aston Martin DB 5	7.0	13.5	6.5
45 DCOE 9	Coventry Climax 1500 GT	5.0	13.5	8.5
45 DCOE 9	Maserati 3500 GT Special	5.0	13.5	8.5
45 DCOE 13	Austin Healey 3000	8.5	15.0	6.5
45 DCOE 14	Alfa Romeo Giulia TI Super GTA	8.5	15.0	6.5
45 DCOE 15/16	BMW 1800 TISA	7.5	14.0	6.5

Note: Dimensions A, B and Stroke are in millimeters.

DCOE INITIAL SETUP & TUNING
Cold Engine:
- If float level has not been set, do it now.
- Check for fuel leaks. If using an electric fuel pump, turn on ignition so it operates and then check all fuel fittings for leaks. If a mechanical pump is used, disconnect high-tension (large) wire from coil, crank engine for several seconds and then check for fuel leaks. Reconnect coil wire.
- Have a helper fully depress accelerator pedal. Verify throttle plates fully open. If not, readjust linkage so they open completely.
- Verify, if necessary, that idle-speed and idle-mixture adjustment screws are at their initial settings. Each idle-mixture screw should be backed off from its seat an equal amount.
- Verify mechanical throttle linkage is synchronized so throttle plates of all carburetors open simultaneously.

Warm Engine:
- Connect a tachometer to engine to monitor idle speed. Start engine and let it warm to operating temperature. This may require resetting idle speed so engine continues to run.
- Use airflow meter, Unisyn or equivalent, to verify each carburetor is flowing the same amount of air. Adjust each as necessary to achieve equal flow between carburetors. Refit air cleaners.
- Adjust idle speed to approximately 850 rpm. Any speed below 1000 rpm is acceptable so long as intermediate or main circuits are not operating and engine continues to run.
- Turn each idle-mixture screw equally to adjust idle mixture so engine idles at maximum rpm. Some amount of experimentation will be required to obtain this setting. Start by turning screws out in equal increments of 1/2 turn—rotate slot in screw's head exactly 180°—until engine rpm begins to decrease. If turning screws out only increases engine rpm, reset screws to initial setting and readjust baseline idle speed so it is as low as possible. Then, repeat idle-mixture adjustment. When engine rpm begins to decrease, turn in idle-mixture screws equally until maximum idle speed is obtained.
- Readjust idle speed to 850 rpm, or as specified in owner's manual or underhood label.
- Disconnect tachometer.

Triple-throat IDA3C.

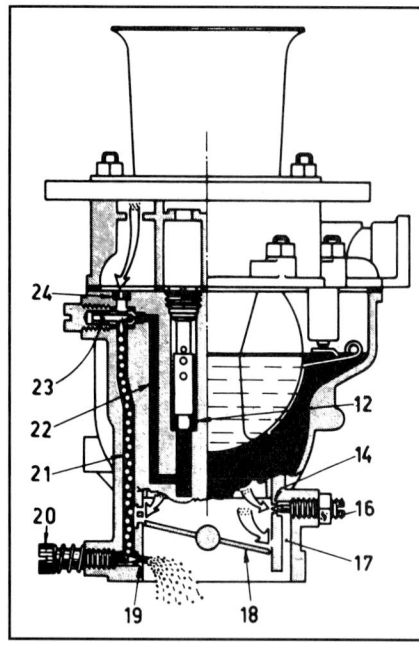

IDA idle circuit uses bush for air-correction jet. Fuel is fed from bottom of well, so carburetor is a monojet. Drawing courtesy Weber.

Progression circuit is typical. Note air-bleed screw at right, which bypasses throttle plate and permits easy balancing of airflow at idle. Drawing courtesy Weber.

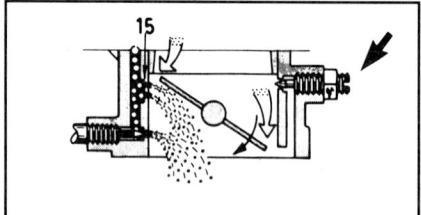

TRIPLE-DOWNDRAFT WEBER: 40 IDA3C

The IDA3C carburetor is used for high-performance applications, primarily on Porsche engines. The *two-venturi* or *two-throat* IDA is very similar in concept to the DCOE series, and can be considered a vertical equal of the horizontal DCOE. That is, the throttles open simultaneously, everything critical to fuel mixing is replaceable, and the accessibility of all jets is very good.

Many of these generalities also apply to the 40 IDA3C; however, most parts are unique to the three-venturi IDA. Almost no components interchange between the two- and three-venturi IDAs.

The IDA3C uses mostly 8mm and 10mm fasteners. Those two wrenches and a screwdriver are about all one needs to field-strip an IDA3C. That should not, however, encourage you to tackle this Weber. This is because of the cars on which the carburetors are fitted. IDA3C Webers are used on high-performance engines. A miscalculation or misadjustment could quickly cause major and expensive engine damage. There is nothing complex about this carburetor or construction. That fact, though, doesn't make it either foolproof or immune to careless repair techniques.

The IDA series is unique among Weber carburetors in that it has no choke circuit. Cold starts with an IDA-equipped engine are eased by pumping the throttle, which causes the accelerator pump to dump raw fuel into the intake manifold. Indeed, the accelerator pump is the only fuel-enrichment device fitted to this carburetor.

The IDA3C carburetor illustrated here is one of the Webers featuring an *air compensating adjustment* for fine-tuning the volume of air delivered at idle. This Weber design supplies some idle air through a passage that is precisely regulated by a needle valve. Thus, on the IDA3C, both fuel and air are regulated at idle by needle valves, and the idle airflow of each throat can be adjusted independently.

The main casting of the IDA series includes either two or three venturis and carries all jets and their drilled passages. The float bowl casting carries only the float assembly.

It's not practical to disassemble this carburetor completely just to service one part. You can do more accidental and expensive harm disassembling an IDA3C than could ever result from normal wear. Spend some extra effort to diagnose a problem to prevent breaking a perfectly good part during disassembly.

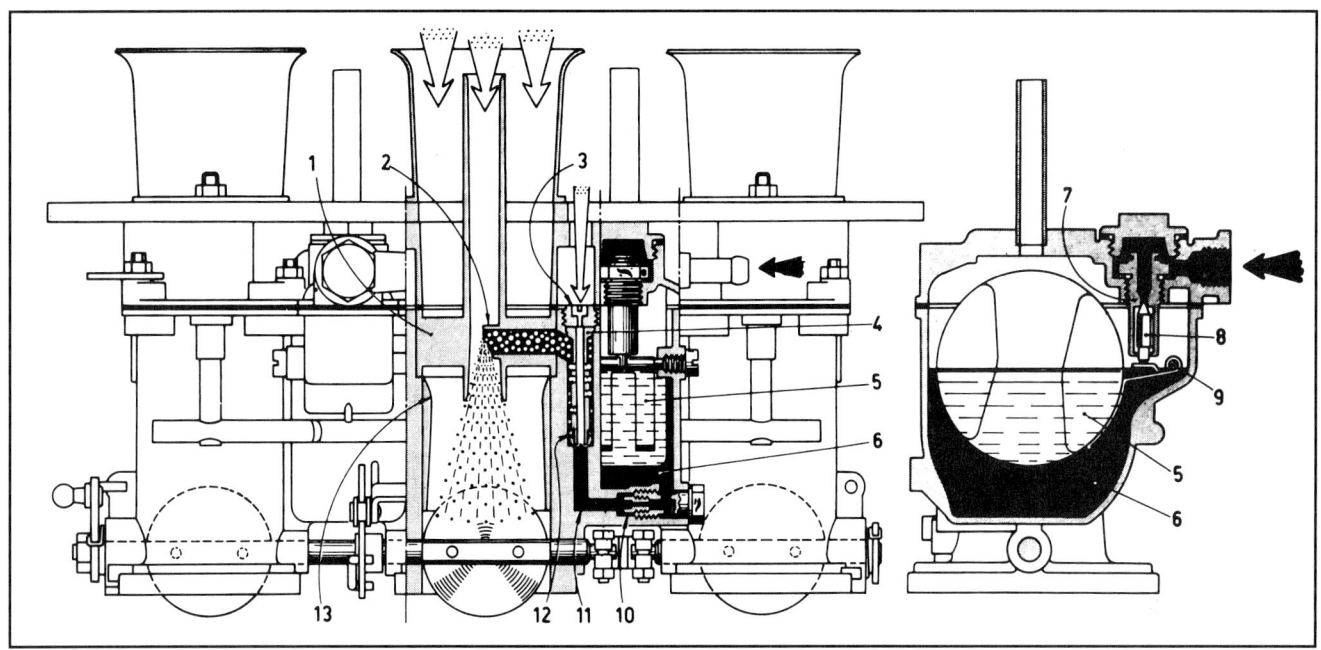

Main circuit is straightforward. Main jet (10) has its own holder low on carburetor float-bowl body. Air-correction circuit is accessible from top of carburetor through holes in float-bowl cover. Drawing courtesy Weber.

The most common wear points are the throttle shaft bearing bores in the carburetor body of triple-throat models. Two-throat carburetors in this series use ball bearings at the outboard ends of the shafts and aren't likely to wear. The triple-throat model has more bearing surface area for the throttle shafts, and so doesn't wear significantly. Weber makes oversize shafts for the carburetors, should the bearing surfaces need to be renewed by reaming oversize. The reaming operation itself is beyond the capability of most amateurs.

Most home-bound repairs are more likely to be extensive cleaning and checking jobs, not overhauls involving reaming seats and passages, or fitting oversize throttle shafts. Beyond a thorough cleaning, setting the float level and replacing rubber parts and gaskets, leave comprehensive Weber overhauls to a Weber specialist.

As with all Webers, removing the throttle plates is the most challenging repair operation. If the shafts don't have to be removed, refurbishing an IDA is straightforward enough that it can probably be performed without removing the carburetor from the car.

The procedure described below is divided into sections and covers the complete disassembly of the carburetor after it has been removed from the engine:

- Removal and disassembly of float bowl cover.
- Removal of components accessible when float bowl cover is off.
- Disassembly of components on outside of carburetor.
- Removal of throttle shafts.

This is not the only order in which assemblies can be removed. Think of the carburetor as a collection of modules, any one of which can be worked on without disturbing the others. Here are the three main modules.

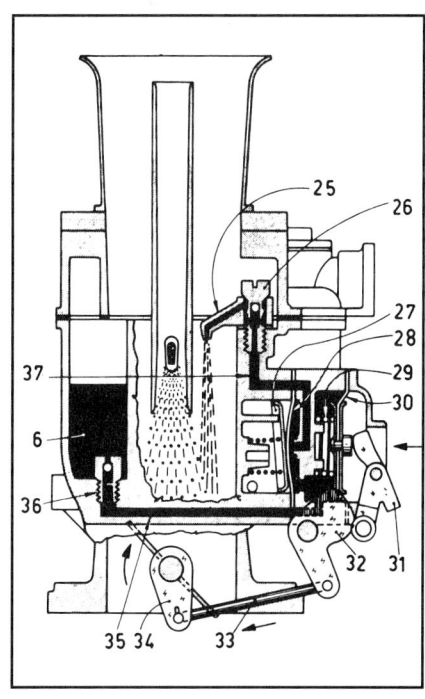

Diaphragm-type accelerator pump is compact, uses two diaphragms to deliver pressurized fuel. Drawing courtesy Weber.

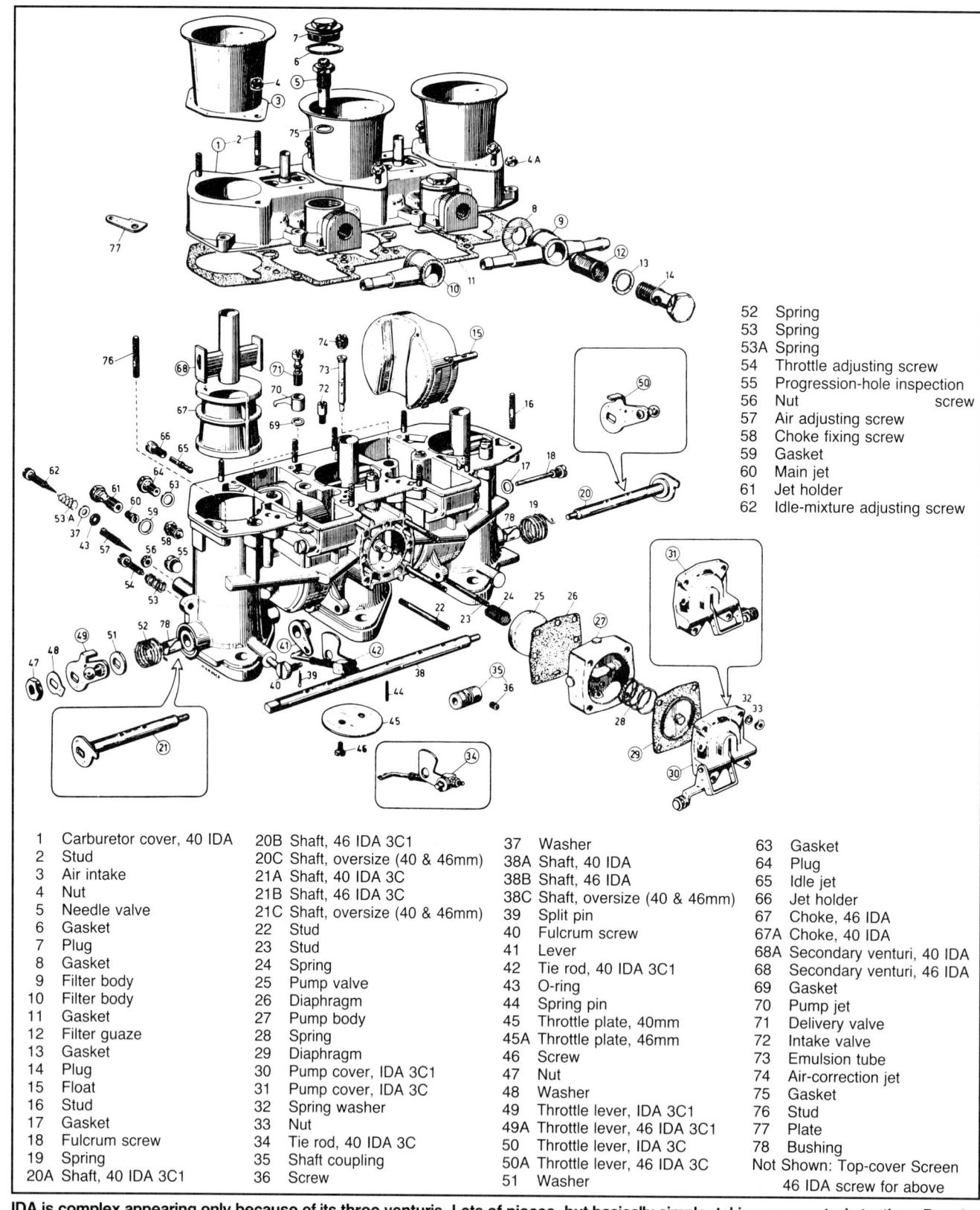

					52	Spring
					53	Spring
					53A	Spring
					54	Throttle adjusting screw
					55	Progression-hole inspection screw
					56	Nut
					57	Air adjusting screw
					58	Choke fixing screw
					59	Gasket
					60	Main jet
					61	Jet holder
					62	Idle-mixture adjusting screw

1	Carburetor cover, 40 IDA	20B	Shaft, 46 IDA 3C1	37	Washer	63	Gasket
2	Stud	20C	Shaft, oversize (40 & 46mm)	38A	Shaft, 40 IDA	64	Plug
3	Air intake	21A	Shaft, 40 IDA 3C	38B	Shaft, 46 IDA	65	Idle jet
4	Nut	21B	Shaft, 46 IDA 3C	38C	Shaft, oversize (40 & 46mm)	66	Jet holder
5	Needle valve	21C	Shaft, oversize (40 & 46mm)	39	Split pin	67	Choke, 46 IDA
6	Gasket	22	Stud	40	Fulcrum screw	67A	Choke, 40 IDA
7	Plug	23	Stud	41	Lever	68A	Secondary venturi, 40 IDA
8	Gasket	24	Spring	42	Tie rod, 40 IDA 3C1	68	Secondary venturi, 46 IDA
9	Filter body	25	Pump valve	43	O-ring	69	Gasket
10	Filter body	26	Diaphragm	44	Spring pin	70	Pump jet
11	Gasket	27	Pump body	45	Throttle plate, 40mm	71	Delivery valve
12	Filter guaze	28	Spring	45A	Throttle plate, 46mm	72	Intake valve
13	Gasket	29	Diaphragm	46	Screw	73	Emulsion tube
14	Plug	30	Pump cover, IDA 3C1	47	Nut	74	Air-correction jet
15	Float	31	Pump cover, IDA 3C	48	Washer	75	Gasket
16	Stud	32	Spring washer	49	Throttle lever, IDA 3C1	76	Stud
17	Gasket	33	Nut	49A	Throttle lever, 46 IDA 3C1	77	Plate
18	Fulcrum screw	34	Tie rod, 40 IDA 3C	50	Throttle lever, IDA 3C	78	Bushing
19	Spring	35	Shaft coupling	50A	Throttle lever, 46 IDA 3C		Not Shown: Top-cover Screen
20A	Shaft, 40 IDA 3C1	36	Screw	51	Washer		46 IDA screw for above

IDA is complex appearing only because of its three venturis. Lots of pieces, but basically simple, taking one venturi at a time. Drawing courtesy Redline, Inc.

- Float bowl cover
  Carries: fuel filter
  Gives access to floats, float needle valve, air-correction jets and emulsion tubes, accelerator-pump delivery valve and venturis.

- Exterior of carburetor
  Main, Intermediate and Idle jets
  Accelerator pump

- Throttle shafts and plates.

The procedure that follows only suggests the order in which assemblies may be removed. Variations are possible, depending on the repair needed.

FUEL FLOW

Fuel flows through the carburetor as shown in the accompanying diagrams.

DISASSEMBLY

See the sidebar, page 45, for information on special Weber tools and their application. If you're going to do a complete overhaul, buy an overhaul kit. If you're doing anything less than a complete overhaul, try to save most of the old gaskets and diaphragms necessary to successfully reassemble the DCOE. If you're trying to save a gasket or diaphragm, work very slowly and carefully to separate it from its mating surface.

Begin the overhaul by cleaning the outside of the carburetor as completely as possible. Scrape off baked-on dirt with a knife or screwdriver, but don't gouge out pieces of the carburetor in an attempt to clean its surfaces. Use an aerosol can of carburetor cleaner to remove varnish and any remaining dirt. Follow the instructions and precautions on the can. Use goggles and avoid breathing the spray.

It's the mark of a professional to make sketches and notes when disassembling a carburetor. Operations such as dismantling a throttle shaft, particularly one spanning three venturis, create lots of parts that must be reinstalled in their correct order. Any diagrams, no matter how rudimentary, are helpful when facing a handful of levers, washers and springs.

Air horns are retained with self-locking nuts. If nuts come free easily, replace them or use lock washers under nuts for reassembly.

Fuel-filter screens surround inlet-union bolts. Clean with compressed air.

The IDA uses many *Nylock* nuts, which use nylon inserts for self-locking. They deteriorate; if the carburetor is used for racing, install new ones during reassembly. If you can turn a Nylock nut with only your fingers, then the self-locking property of the nut isn't sufficient. Replace it. Nylock nuts are typically fitted without washers.

Float Bowl Cover—The IDA3C is different from other Webers in that the floats are attached to the main body and not removed with the float bowl cover. Begin disassembly by removing the air-intake horns. Unhook the throttle-lever return spring and pull the intermediate hose from between the fuel inlet unions.

Remove the fuel inlet unions and fuel filter screens from around the union bolts. Save the red fiber sealing washers for reuse.

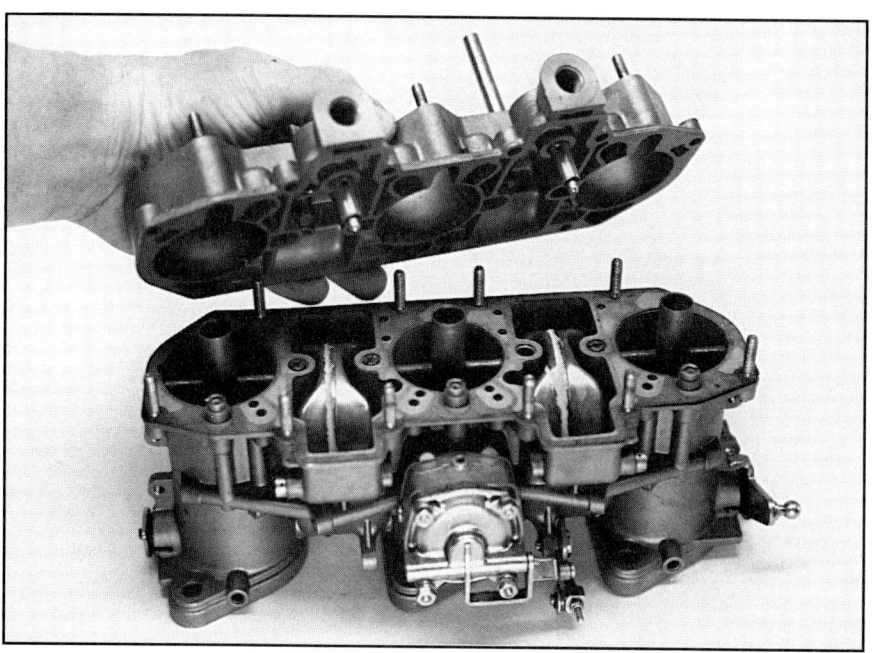
IDA 3C float-bowl cover is handfull. Needle-valve housings are visible; remove with 14mm socket.

IDA has externally-mounted float-pivot pins. Lock wire assures they stay put.

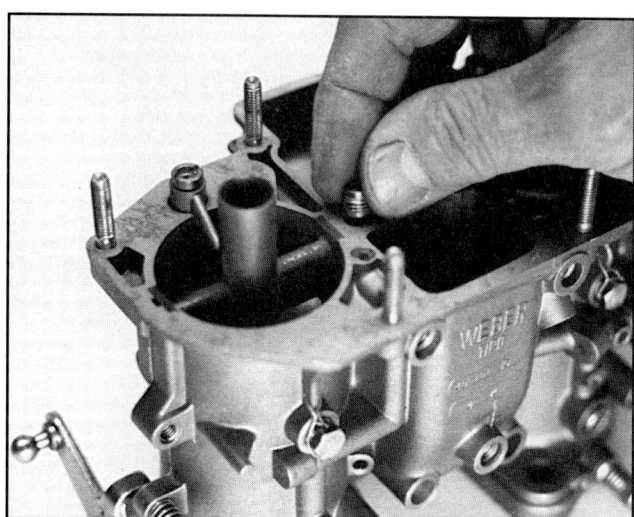
Air-correction jet sits between float bowl and venturi. Be careful not to damage jet surface with screwdriver during removal.

Emulsion tube is trapped in well below air-correction jet. It should drop out. If it doesn't, run in a small tap just enough to get needed grip to extract tube.

Unscrew the needle valve covers. An aluminum washer beneath each cover may appear to be part of the casting. Be sure to remove it and store it for reuse.

Unscrew the ten 8mm Nylock float cover retaining nuts and lift the cover free. Remember that the accelerator return-spring tab is secured by one of the nuts. Carefully remove the cover and gasket. Both the studs and gasket are long, so work the gasket off gently.

Use a 14mm socket to loosen the needle valve housings from the float bowl cover. Unscrew the housings with your fingers from underneath once the nut is loose. Save the washer beneath each housing.

Remove the locking—safety—wire from the float-fulcrum pin screw heads and remove the pins and washers. Then remove the floats from the main body.

Accelerator-pump delivery jet sits high, holds nozzle in place. Jets and nozzles are interchangeable between venturis. Cast-in passages on body leave no doubt regarding fuel path.

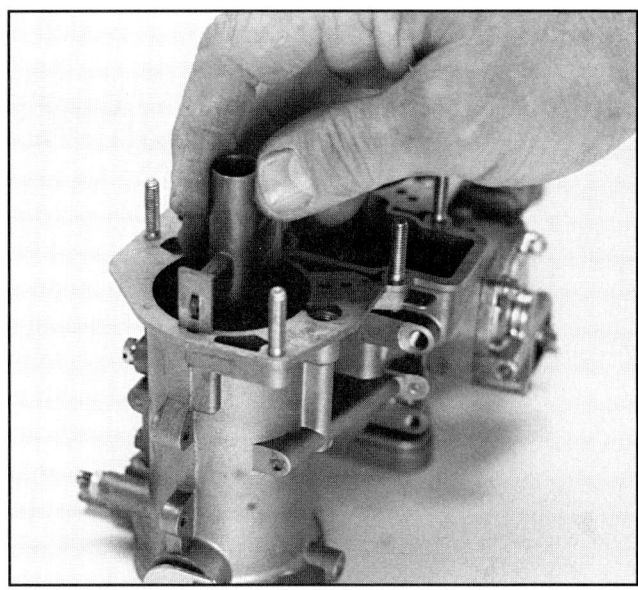

Remove auxiliary venturis by pulling up on long center tube.

Jets & Emulsion Tubes—Unscrew the float bowl drain plugs and washers. Unscrew and remove the main jet holders and washers, then unscrew the jets from the holders. To unscrew the jet, use a screwdriver with a blade exactly the size as the slot in the main jet. Use a 10mm wrench to steady the holder.

Unscrew the idle jet holders with their rubber O-rings and pull the idle jets out of the holders.

Unscrew the air-correction jets and remove the emulsion tubes beneath them. You may have to use a 3mm tap to remove the tubes if they're stuck. If you use a tap, cut only enough threads to extract the tube. Then blow the chips out of the bore in the tube with compressed air. Check that there is no debris inside.

Accelerator-Pump Valve & Jet—Unscrew the accelerator-pump inlet valve from the bottom of the float chamber. Check it for one-way operation by blowing through both ends. You should be able to blow only one way. Remove the accelerator-pump delivery jet, nozzle and washer beneath the nozzle. Set each aside as a separate assembly.

Auxiliary Venturis—Pull the auxiliary venturis out. Remember that the taller section of the tube goes to the top. Because the mounting flanges are slightly tapered, there's only one way you can reinstall the venturi. It should slip right in. Mark each auxiliary venturi so you can install it in the same bore at reassembly; don't mix them up.

It may be hard to get an auxiliary venturi out. But don't use pliers to grasp the venturi. You may break it. Instead, use a wood dowel that fits beneath—past the open throttle valves—and use it to tap the venturi out.

If this method fails, skip the removal of the auxiliary venturi here, and try again after the throttle valves and shafts have been removed. Remember, there is a Weber tool for removing stuck auxiliary venturis.

Accelerator Pump—The accelerator pump assembly, on the outside of the carburetor, has two diaphragms with a spacer between. The entire assembly is held together by four nuts.

Unscrew the accelerator-pump retaining nuts in increments to evenly release pressure from the diaphragm. Remove the nuts and spring washers, then remove

Four nuts hold accelerator pump to body of carburetor. Incrementally loosen all four to avoid stressing diaphragm inside.

the accelerator pump cover with the diaphragm and catch the large return spring.

Tap on the spacer to loosen it, then slip it off the studs. Work a screwdriver behind the rubber diaphragm and ease it off the studs. A return spring and plate will fall out from behind the diaphragm.

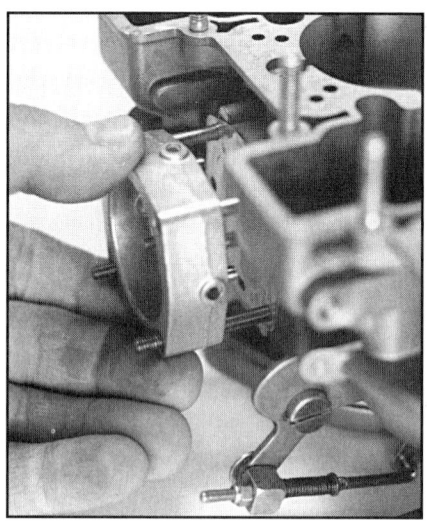

Remove spacer. Another diaphragm fits between spacer and carburetor body. Plate and spring are trapped behind this diaphragm.

Air-compensation screws should be removed with locknut in nearly original position. Loosen nut only enough to free screw for removal.

Idle-mixture screws should be seated first, noting number of turns in, then removed. On reassembly, run in to seat screw, then turn out same number of turns noted during disassembly.

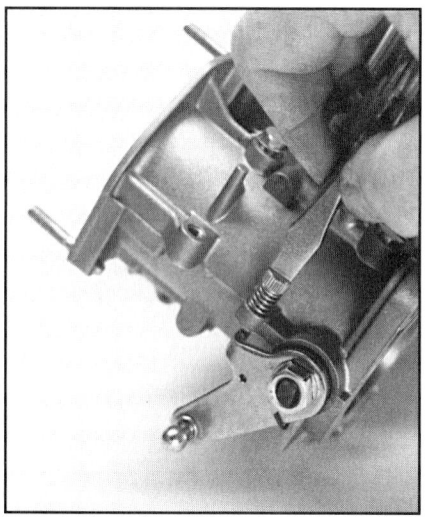

Idle-speed adjustment screw need not be removed unless carburetor is undergoing major overhaul, including removal of throttle plates.

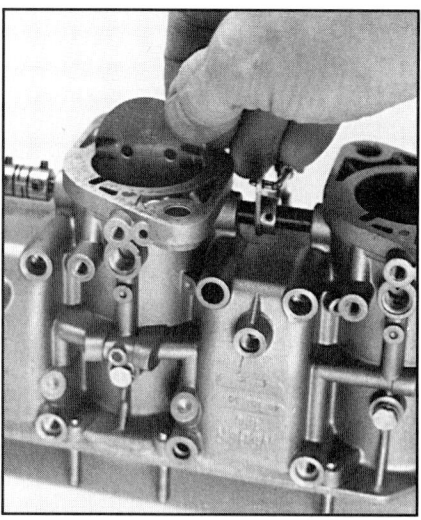

Removing throttle plate is not a good idea, especially on expensive carburetors such as IDA 3. Pitfalls are many, and new parts, including throttle shaft, may be needed. Unless there is an overwhelming need to remove plates, leave them in place.

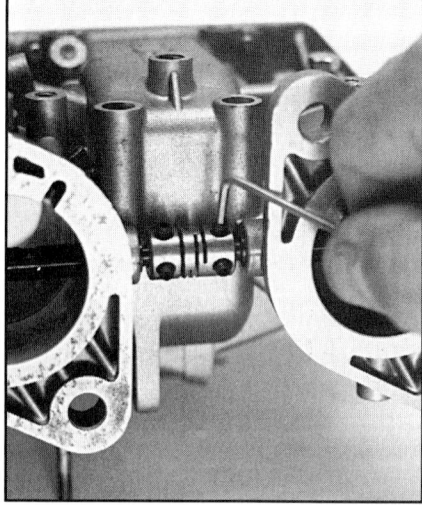

Throttle plates are synchronized with these couplings. Loosen Allen screw, don't remove it. Short throttle shaft is to right as viewed.

Air Compensation Screws—Loosen the locknuts about 1/8 turn, then use a screwdriver to remove the air compensation adjustment screws. If you don't move the locknut, you'll have a starting point to readjust.

Idle Mixture Screws—Seat the idle mixture adjusting screws lightly, recording the number of turns required. Then remove the idle mixture assembly, which will come out as a single unit: idle screw, spring, O-ring and protective washer. Remove the idle speed adjustment screw and spring. Remove the progression-hole inspection plugs. They are just above the idle mixture screws.

Throttle Valves—Don't remove the throttle valves and shaft unless **absolutely necessary**. The chances of making costly errors and ruining parts are high. So, what's absolutely necessary? These

 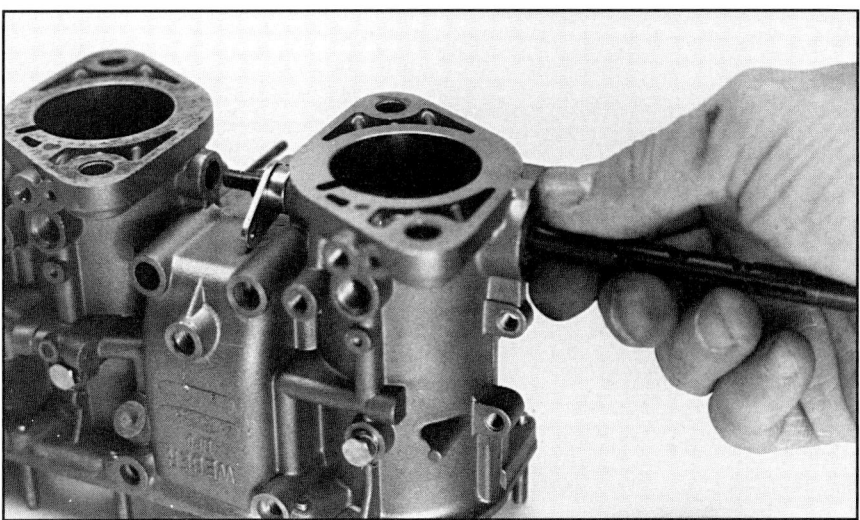

Accelerator-pump-cam lever is attached through linkage to throttle shaft. Lever has pivot screw secured with cotter pin. Remove pin with needle-nose pliers; refit new pin on reassembly.

After accelerator-pump lever is free, pull out long throttle shaft.

conditons: If a throttle valve is binding and can't be freed, or its action can't be improved with cleaner, or the shaft is bent or worn. Consider, too, you must also remove the auxiliary venturi and venturi to reinstall the throttle valves.

A worn throttle shaft will produce an erratic idle speed or inconsistent return to a stable idle. If faced with these conditions, then consider replacing the complete unit or sending it to a Weber specialist to do the repair. The IDA3C is an expensive carburetor so reconsider tackling such an extensive repair. I've included removal and assembly for those with a large cash reserve, no fear, steady nerves, and ample time.

First extract the throttle valve retaining screws. The ends of these screws have been staked to keep them from vibrating loose and falling into the engine. Sometimes, it's possible to remove them without special effort. Work very carefully, for there are only a few threads in the throttle shaft to hold the screws.

Try to unscrew them by working them out in increments of 1/2 turn, alternated by turning them in 1/4 turn. If this approach does not work, simply grind off the staked portion of the ends using a hand grinder. Take care not to damage the throttle bore. Replace the screws at reassembly.

Rotate the throttle shaft and withdraw the throttle valves. Mark each so it can be returned to its own bore exactly as it was removed.

Removing the throttle valve attaching screws invariably raises burrs on the throttle shaft. Smooth the shafts with a file to remove any burrs. Otherwise, they will damage the Teflon bushing at the end of the shaft when it's removed.

Throttle Shafts—Loosen, but don't remove, the set screws that attach the adjusting coupling to the throttle shafts and withdraw the *short* throttle shaft.

Remove the return spring from the short throttle shaft and inspect the Teflon bushing for damage. The bushing usually comes out with the shaft. If it doesn't, remove it from the body of the carburetor with a small hook.

Remove the cotter key that secures the pivot screw for the accelerator-pump cam lever. Undo the pivot screw and take off the accelerator-pump cam, unhooking the linkage to the *long* throttle-shaft bellcrank.

Drive out the split pin that attaches the accelerator-pump actuating lever to the long throttle shaft. Remove it.

If there is no reason to further disassemble the long throttle shaft, leave it intact. If you want to repair it, then remove the linkage from the long throttle shaft. Bend the lock tab and undo the 12mm nut. Remove: nut, lock tab, bellcrank, spring, washer and Teflon bushing. Store these parts.

Venturis—To remove them, detach the safety wire from each retaining screw, remove the screws completely and withdraw each venturi. Mark the location of each, then knock it free with a wood dowel from below if it doesn't drop out. Remember that the fat part of the venturi with the number on it goes up.

Main venturi is held in with lock screw. Loosen screw and venturi should slide out easily.

INSPECTION

Cleaning—See the cleaning section, page 51, of the IMPE part of this chapter for tips on cleaning the carburetor.

Inspect the drilled passages in an IDA as you clean them with an aerosol spray or ear syringe full of solvent. Wear goggles when you do this. And work in a well-ventilated area away from an open flame or pilot lights.

Aerosol cleaner cans come with a plastic tube that inserts into the spray nozzle. Use the tube to probe passages while spraying cleaner. It's unusual for passages to be completely blocked. Dirt usually blocks the jets before it clogs the passages. A more common problem to look for is the loss of plugs used to block the ends of drilled passages.

The high cost of an IDA3C is a good reason to avoid doing extensive or heroic repair procedures. Send such repairs to a Weber specialist.

Float Bowl Cover—The gasket sealing the float bowl cover to the main casting should show that the sealing surfaces of the two castings are parallel. Replace the gasket if it's torn or brittle. Check the float bowl cover for broken flanges, cracks or other physical damage. Check the threads for the fuel filter plug.

Fuel Filter—Check the fuel filter for any damage to the screen. Replace the screen if it is torn. Blow it clean with compressed air to remove any dirt.

Floats—They shouldn't show any evidence of leaking. Shake the assembly near your ear and listen for gasoline sloshing around inside. If the floats leaked, replace them. The weight in grams of each float is stamped on the arm of each IDA float. If in doubt about leakage, weigh the float. If it is heavier than indicated, it has fuel inside. Replace it.

Needle Valve—The IDA3C float needle valve can't be disassembled because it is crimped inside the housing. The valve should move freely within the housing and there should be no visible wear on the ball protruding from the valve. The ball is crimped into the end of the needle valve assembly and held with a weak spring. You should be able to press the ball into the needle valve easily. If you detect any binding at all of the ball or the needle valve, replace the assembly.

Auxiliary Venturis—If you removed them, blow though the passage and remove any burrs or dirt on their surfaces.

Main Casting—Check for damage to the main casting. Use a straightedge to check the carburetor base for warpage.

Jets & Emulsion Tubes—Blow through the main jets, air-correction jets, emulsion tubes and accelerator-pump jet to clean them. Never use a wire or drill to probe or clean the jets. Blow through the accelerator-pump delivery jet to verify its one-way operation. Check each jet opening by holding it up to a light. Check the diaphragms of the accelerator pump for cracks or tears.

Throttle Shafts—These should be straight, with no visible wear where they rotate in the nylon seals. Roll them on a flat surface to check straightness. Double-check the threads at the end of the throttle shaft and the threads where the throttle valves attach. The throttle valves should have smooth edges and be flat. All the springs that slide onto the throttle shafts should be strong and not deformed. Replace the nylon shaft seals whether or not they show any wear.

REASSEMBLY

The disassembly of the IDA3C is, in itself, a thorough lesson in reassembly. After all the components have been inspected and damaged parts replaced, you can begin reassembling any of the major components first. Start with either the float bowl cover or main casting. The following notes give the most important points to remember in reassembly.

Throttle Shafts—There are two dangers in reassembling the throttle shafts: bending the shafts and not correctly positioning their valves. The holes in the throttle valves are oversize so they can be positioned for a perfect seat in the bores. The auxiliary venturis should be removed to give free access to the attaching screws during this operation.

Assemble parts to the long throttle shaft first. Working outward from the

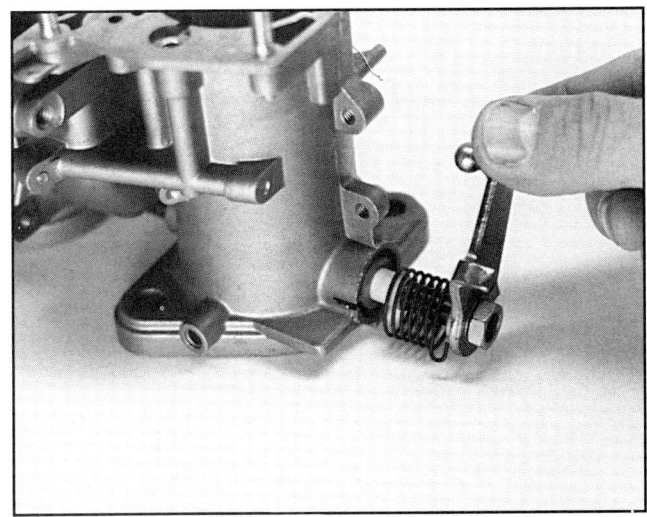

Make sure you put the two throttle shafts on the correct ends of the carburetor! Long throttle shaft is started into bore with bushing, spring, spacer and so forth in place as shown.

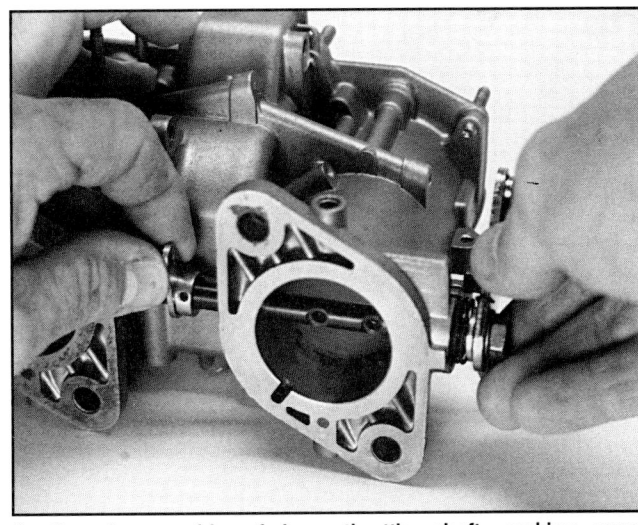

Continued assembly of long throttle shaft, making sure accelerator-pump bellcrank is correctly positioned. Split pin secures bellcrank to shaft. Return spring should be engaged in slot on body of carburetor as shaft slides in. Other end is slipped over arm after shaft is almost home.

innermost part, the order is:

- Teflon bushing—fit it inside the carburetor body
- Spring
- Spacer
- Throttle bellcrank
- Locking tab
- Nut

While feeding the shaft into the body, slip the accelerator-pump arm onto the shaft. Double-check the orientation of both the accelerator-pump arm and throttle bellcrank. Pin the accelerator-pump arm to the shaft using a new split pin. Then, slip on the coupling to the throttle shaft. Don't tighten the set screws yet.

To assemble the throttle return spring, place its straight end in the body of the casting, then slip the throttle bellcrank halfway onto the long throttle shaft. Hook the curved end of the spring over the bellcrank. Wind the bellcrank counterclockwise to tension the spring, then press it home so it is held in position against the idle speed stop. With the bellcrank seated against the idle stop, tighten the nut at the end of the throttle shaft, then bend over the lock washer.

The short throttle shaft has its own Teflon bushing and spring that should be installed using the same technique as used for the long throttle shaft. Place the Teflon bushing in the carburetor body, then put the return spring on the short shaft. Feed the shaft into the carburetor body so that it connects with the long shaft's coupling. Press on both the long shaft and short shaft to make sure they're fully seated in the carburetor body. Then position the coupling so it is equally spaced between the shoulders of the short and long throttle shafts. Don't tighten the set screws. Their final adjustment is made after the throttle valves have been installed.

Throttle Valves—Notice that the edges of the throttle valves are beveled. Verify that the beveled surface fits flat against the venturi bore. Most valves have a degree value stamped on the side that faces the carburetor base.

Throttle-valve screws can vibrate free and drop into the engine, causing severe damage. Use fuel-resistant Loctite on the

Slip short throttle shaft in place and engage with coupling on end of long shaft. Drill for return spring is same as for long shaft.

Secure throttle-shaft screws. Ends of screws must be staked to keep them from loosening and falling into engine.

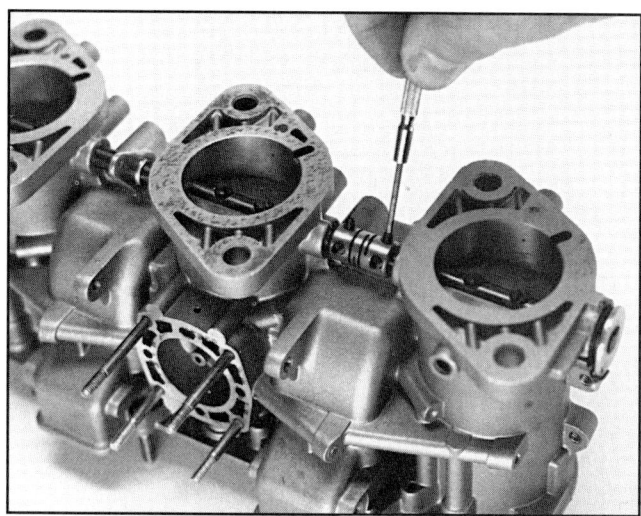
Synchronize long and short throttle shafts after throttle valves are in place. This is critical. Work slowly and carefully. Don't tighten Allen screws until you're sure shafts are in synch.

threads and stake the screw ends. You can deform the screw ends with a chisel or punch only if you support them correctly so no force is transmitted to the throttle shafts. Otherwise, you'll bend the shafts and have to replace them.

Slip one throttle valve into its slot, apply the Loctite, then tighten its attaching screws. Replace the next valve, positioning it and fitting its attaching screws before proceeding to the final valve. Be careful to replace each valve in its original position and double-check that the valves seat themselves in their bores perfectly. If you have any trouble, the valves are probably not positioned correctly.

Double-check your work by holding the carburetor up to a bright light and look around the circumference of the valve. You should be able to achieve an almost light-tight seal of the valve to the bore. If there is a crescent of light, it should be concentric with the valve.

Tighten the screws securely and recheck your work. The valves are held in place by the tightness of the screws. Stake the ends of the screws so they'll stay tight.

Stake the screw ends with a sharp chisel while the screw head is supported so you don't bend the throttle shafts. Set a 1/4-in. drive, 1/2-in. socket on a steel or concrete surface with the square drive end facing up. Place the carburetor on top of the socket so the head of a throttle valve screw rests against the smooth surface of the socket. Then, place a sharp chisel down the venturi and position it on the end of the screw being staked.

The valves won't allow the throttle shaft to rotate so you can hit the screws dead-center, but the angle is so small that it won't matter. Try to deform the screw with a single sharp blow directed as close to center with the screw as possible.

You can also secure the screws with Vise-Grip pliers. Reach inside the venturi with a small pair of them to squish the threads on the exposed screw end. This will deform the thread and keep the screw from backing out.

Synchronizing Throttle Valves— Now, verify that both throttle shafts are in the fully closed position. The idle speed adjusting screw isn't fitted yet, so the idle stop is holding the throttle valves closed on the long throttle shaft. Use finger-pressure to ensure that the throttle valve on the short shaft is closed. Now tighten the connection between the two shafts. All three throttle valves must close simultaneously and snugly. Open the throttle, then close it to verify that all the valves operate in unison.

Readjust the valve on the short shaft if necessary by loosening, then retightening the coupling. If the set screws of the coupling have significantly embossed or dented the shaft, then rotate the coupling so the adjustment won't be upset by the set screws seeking their old position.

Venturis—Refit the venturis and auxiliary venturis in their original bores. Note that the number cast into the venturi is on line with the set-screw seat. Lock the venturi in place with the set screw, then tighten the set-screw locknut. Wire the screws in place using new safety wire. Use only wire intended for safety-wiring: Don't use a strand of copper electrical wire; use stainless-steel safety wire.

Accelerator-Pump Cam—Refit the accelerator-pump cam assembly to the throttle-shaft bellcrank. Smear a little grease on the bearing surface of the cam attaching screw, then attach the cam to the carburetor body. Fit a new cotter key

Refit venturi. Notice how depression in venturi's outer diameter aligns with locking screw on outside of carburetor body. Be sure screw engages depression before tightening it.

Hook accelerator-pump link through bellcrank on throttle shaft, then position cam at carburetor body. Attach with shoulder screw as shown.

to secure the screw. Test the assembly to verify that it moves freely.

Inspection Plugs—Smear a very light coat of sealer on the threads of the progression-hole inspection plugs and refit them. Install the plugs as tightly as you can manage with a large screwdriver.

Idle Adjustment Screw—Assemble the spring to the idle adjustment screw and reinstall it so it just begins to open the throttle plates.

Idle Mixture Screw—Refit the idle mixture adjusting screw assemblies: adjusting screw, springs, O-rings and protective washers. Seat the screw lightly, then back it out the same number of turns you noted on disassembly. Try 1-1/2 turns out as an initial richness adjustment if you have no baseline.

Air Compensation Screws—Refit the air compensation adjustment screws fingertight, then use a wrench to tighten the locknuts approximately 1/8 turn.

Accelerator Pump—Fit the small return spring and plate over the tower inside the accelerator pump casting on the side of the carburetor. Place the flat diaphragm over the plate, then fit the spacer over the diaphragm.

Seat idle-mixture screw, then back out same number of turns as noted during disassembly. Values among the three mixture screws may vary.

Accelerator-pump spacer covers diaphragm, spring. Another diaphragm goes on top.

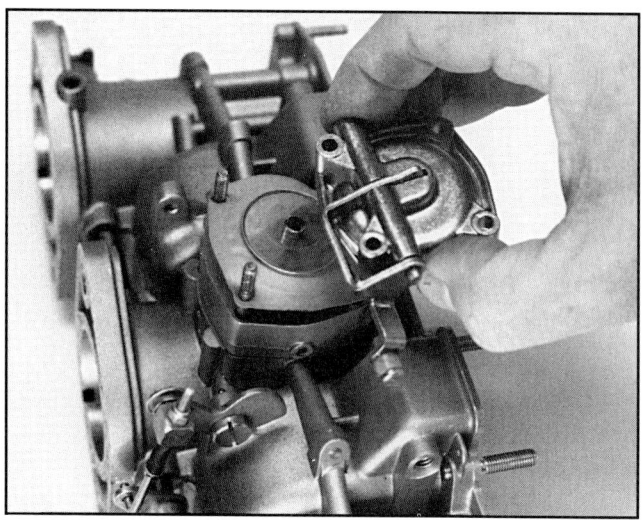
Pump cover is placed over second diaphragm. Note small dab of grease on center boss of metal diaphragm support.

The spacer is a one-way fit, with the leg of the cast-in T pointing to the top of the carburetor. The large return spring is positioned by three small bosses inside the spacer. Put it in place, then cover it with the main diaphragm. One side of the diaphragm has a metal nipple. Smear a very small amount of grease on the nipple and fit the diaphragm to the spacer with the nipple facing out.

Cover the diaphragm with the outer accelerator pump casting, making sure it seats evenly on the diaphragm. Fit the lock washers, then tighten the four attaching screws in small increments until snug.

Reassemble the accelerator-pump delivery nozzles onto the delivery jets, fit a washer under each and screw the jets in place. Refit the accelerator pump inlet valve to the bottom of the float chamber.

Jets & Emulsion Tubes—Drop the air-correction jet emulsion tubes in place, then refit the air-correction jets. *Note:* There is one blank passage near the center venturi which looks like it will accept an air-correction jet. It won't. This passage can be identified because it doesn't have any threads for an air-correction jet.

Fit the rubber O-rings to the idle jets and reinstall the jets, then screw the assemblies into the carburetor body. *Note:* As with the air-correction jet, there is a blank passage that isn't threaded. Don't try to fit an idle jet into the blank passage.

Screw each main jet carefully into its holder, then slip a washer onto the assembly. Reinstall the main jet assemblies into the carburetor body.

Floats—Reassemble both float bowl plugs with their washers, then screw each plug into the carburetor body. Place the floats into the body of the carburetor. Slip a washer on each fulcrum pin, then fit the pin, slipping the float onto its pin. Tighten the pin, then use *new safety wire* to secure the pins. Stainless-steel wire is best. These pins must be secured. If the wire were to break and the fulcrum worked free, the engine would flood with fuel.

Float Bowl Cover—Begin reassembling the float bowl cover by reinstalling the float needle valve assemblies. Slip washers onto the valve assemblies, then screw each assembly into the cover and tighten.

Fit the float bowl cover gasket very carefully over the studs, then put the float bowl cover in place. Put the tab for the throttle return spring over its proper stud, then install the Nylock nuts for the float bowl cover. Tighten the nuts incrementally, working around the circumference of the cover.

Throttle Return Spring—Fit the throttle return spring with the wound section toward the top.

Needle Valve Covers—Install the needle valve covers and washers.

Air Horns—Install the air horns.

Fuel Filters—Reinstall the fuel filter screens, unions and union bolts with their washers, then refit the intermediate fuel delivery hose.

Preliminary Tuning—After the carburetor is reinstalled on the car, warm up the engine and adjust the idle mixture adjustment screw and idle speed screws to obtain the slowest possible engine speed at idle. Adjust each screw the same amount for each carburetor. This is a baseline rough adjustment.

Next, check the airflow at idle for each throat using a *Unisyn* or *Synchrometer*. These are available from Weber dealers. The throat with the highest flow is the

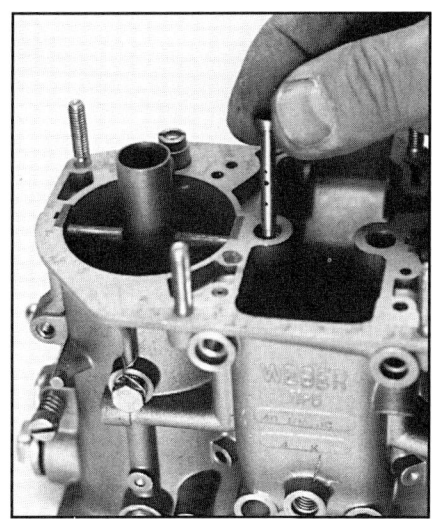

Emulsion-tube replacement: Tubes may be interchanged, because all should have the same values. Tubes are capped with air-correction jet.

Refitting floats: Hold floats while fitting pivot pin from outside of carburetor body.

standard. Adjust the other two throats to match it, using the idle air-bleed adjustment screw.

Now, readjust the idle mixture screws to give the fastest idle without changing the idle speed screw. Next, use the idle speed adjustment screw to set the engine at its specified idle speed. Once again, check airflow in each throat and repeat the air bleed and idle richness adjustment procedure for final tune.

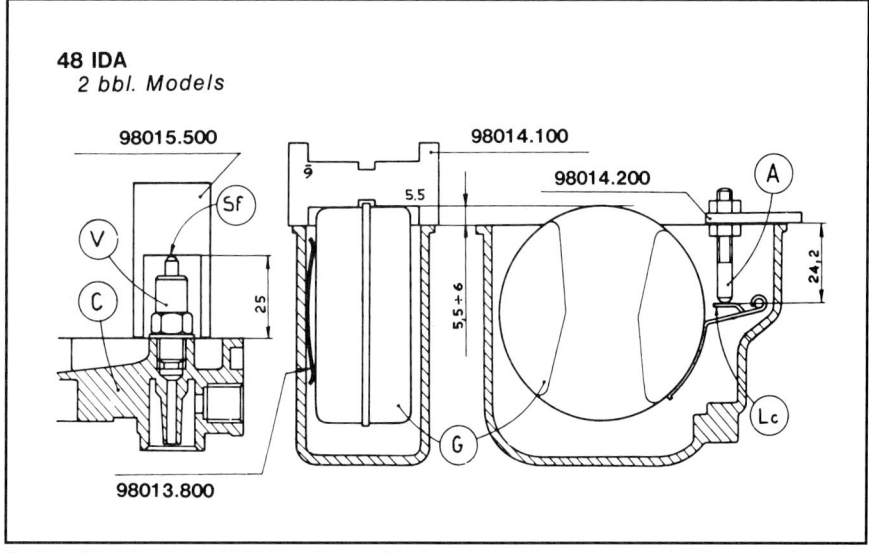

Setting float level on 48 IDA series—with two venturis requires fabrication, purchase or rental of special tools to measure dimensions (in mm). They are available from Weber specialists. One tool measures height of needle valve from body, left drawing. A second is used to set depression of float in bowl, at center, a third is used to measure height of float above bowl face with second tool in place, at right. A spring (center drawing) holds float in position during measurements. Drawing courtesy Redline, Inc.

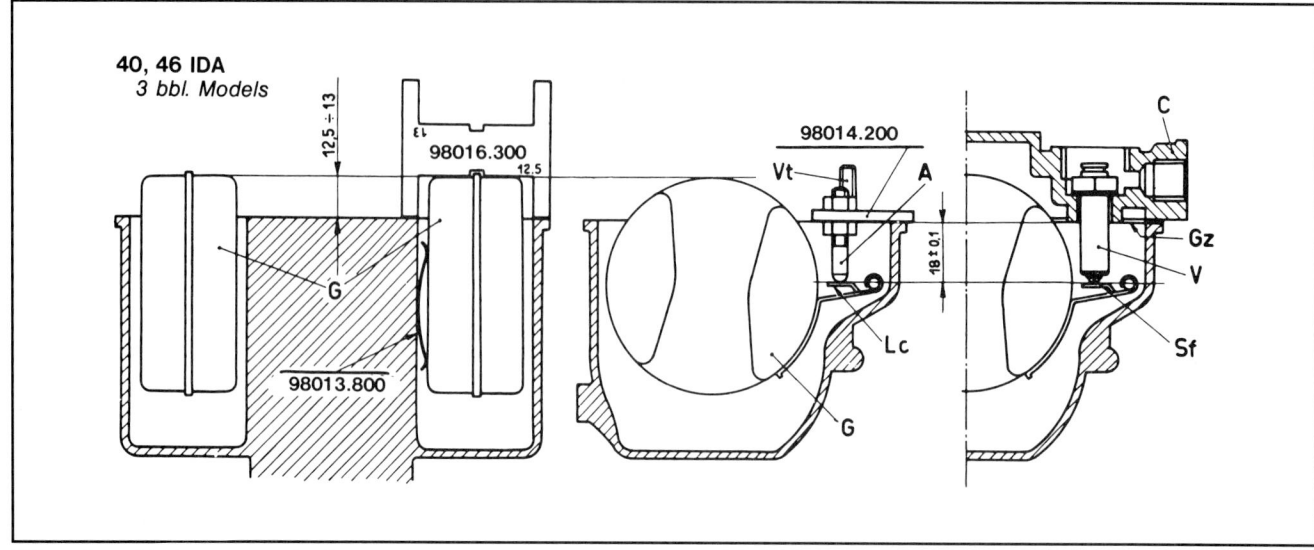

IDA series with three venturis also requires special tools to set float levels. Measurement value between twin- and triple-throat IDA models are different. Drawing courtesy Redline, Inc.

IDA3C INITIAL SETUP & TUNING
Cold Engine:
- If float level has not been set, do it now.
- Check for fuel leaks. If using an electric fuel pump, turn on ignition so it operates and then check all fuel fittings for leaks. If a mechanical pump is used, disconnect high-tension (large) wire from coil, crank engine for several seconds and then check for fuel leaks. Reconnect coil wire.
- Have a helper fully depress accelerator pedal. Verify throttle plates fully open. If not, readjust linkage so they open completely.
- Verify, if necessary, that idle-speed and idle-mixture adjustment screws are at their initial settings. Each idle-mixture screw should be backed off from its seat an equal amount.
- Verify mechanical throttle linkage is synchronized so throttle plates of all carburetors open simultaneously.

Warm Engine:
- Connect a tachometer to engine to monitor idle speed. Start engine and let it warm to operating temperature. This may require resetting idle speed so engine continues to run.
- Use airflow meter, Unisyn or equivalent, to verify each carburetor is flowing same amount of air. Adjust each as necessary to achieve equal flows between carburetors.
- Adjust idle speed to approximately 850 rpm. Any speed below 1000 rpm is acceptable so long as intermediate or main circuits are not operating and engine continues to run.
- Turn each idle-mixture adjustment screw equally to adjust idle-mixture richness so engine idles at maximum rpm. Some amount of experimentation will be required to obtain this setting. Start by turning screws out in equal increments of 1/2 turn—rotate slot in screw's head exactly 180°—until engine rpm begins to decrease. If turning screws out only increases engine rpm, reset screws to initial setting and readjust baseline idle speed so it is as low as possible. Then, repeat idle-mixture adjustment. When engine rpm begins to decrease, turn in idle-mixture screws equally until maximum idle speed is obtained.
- Readjust idle speed to 850 rpm, or as specified in owner's manual or underhood label.
- Disconnect tachometer.
- Using the airflow meter, locate venturi that flows largest volume of air. This is baseline measurement. Use air-bypass adjusting screws to set remaining venturi flow rates so they equal baseline measurement.

6
Modifications

This finely modified 1966 Shelby GT 350 Mustang sports four 48 IDA downdraft Webers. This setup has become a popular choice among Ford enthusiasts. (Ron Sessions photo).

Putting a Weber on an engine is only one of a host of modifications you can make to get more power from your engine. Here I survey and explain some other engine modifications that should work for your application.

There are stories of engine modifications that turn out unpredictably, such as cars that should be able to pull whole buildings—or, at least the bankroll that went into them—but can hardly move their own bodywork; or, evil-running monsters that won't idle, lurch unpredictably at mid-throttle, and only begin to develop power at engine speeds just beyond redline. This section should help you avoid such extremes.

From the outside looking in on an engine, there appear to be many changes you can make to increase engine power: larger carburetor(s), manifold, different camshaft profile, exhaust headers, larger air cleaner(s) and so forth.

But, from the inside of the engine looking out, there is really one fundamental change required to increase power—to increase the pressure that bears on the piston after the sparkplug fires. This *average* pressure is Brake Mean Effective Pressure (BMEP).

An engine modification must increase BMEP in some way if it is to increase engine power. Unless changes, from carburetion to camshaft(s) to exhaust header(s), help improve the internal BMEP, the engine won't act any differently. There are many ways to improve BMEP, and a Weber can contribute in most cases.

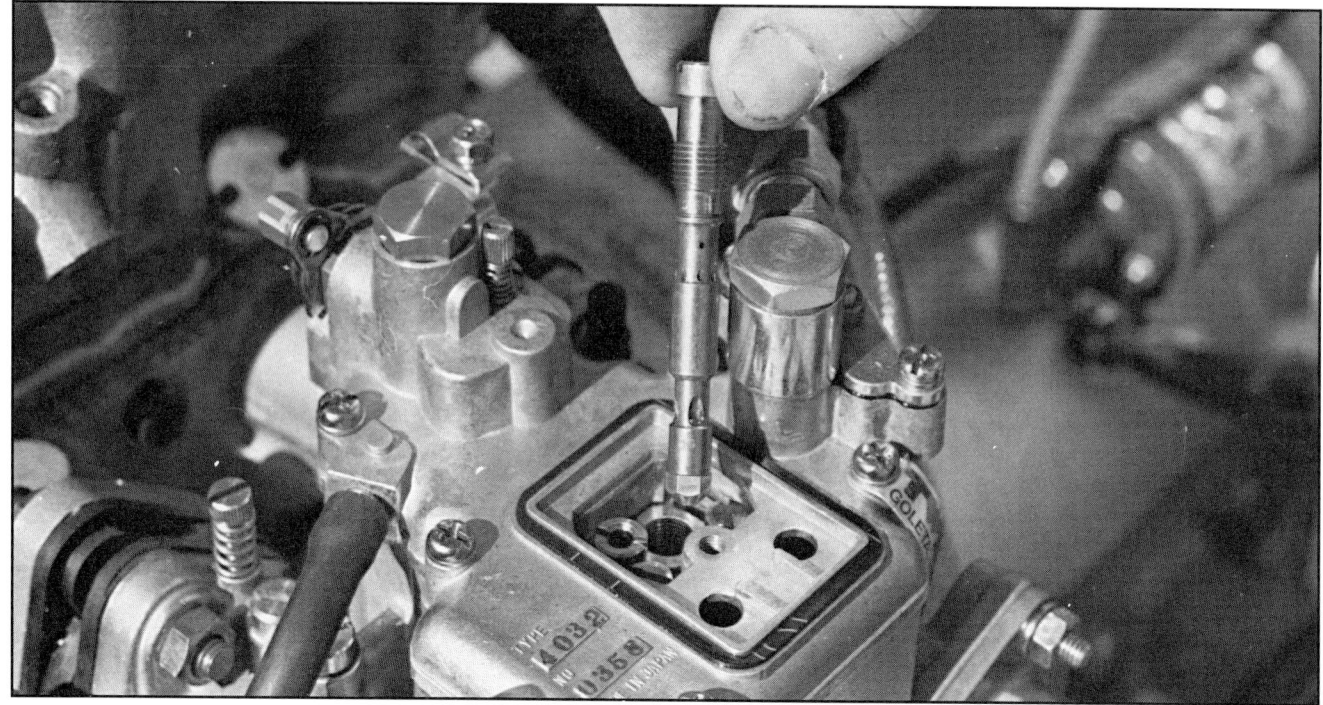

One advantage of performance tuning Webers: their main metering circuits are easily altered. SK Racing carburetor shown here uses standard Weber jets and emulsion tubes. (Photo by Glen Grissom, courtesy MSD Ignition.)

INCREASE COMPRESSION RATIO (CR)

When the air/fuel mixture is compressed, it gets hotter, and the more it's compressed, the hotter it gets. So, the intake-charge temperature is raised during the compression stroke. And, its final burning temperature is raised, too. As a result, the mixture burns more quickly and ends up producing a higher temperature and higher BMEP. Compressing the mixture also improves the charge density and promotes more uniform and efficient burning. Consequently, shaving off a few thousandths from the cylinder head is one of the easiest ways to increase BMEP.

But increasing compression ratio (CR) isn't without problems. First, some exhaust emissions may exceed legal limits; specifically, oxides of nitrogen (NOx) rise; and HC emissions also rise somewhat. Increased NOx emissions result from the higher combustion temperatures that occur when CR is increased.

The CR of engines have decreased from a high of 11:1 during the late '60s to about 9:1 in the mid-'80s. The *octane rating*—antiknock value—of fuel has also fallen, not because the lower-compression engines could survive on it, but because tetraethyl lead was eliminated as an octane booster.

Adding tetraethyl lead to gasoline causes it to burn more slowly, improving its anti-knock—*detonation*—characteristics and, thus, its octane rating.

Detonation, or *knock*, is an explosion in a combustion chamber. Normal ignition of the air/fuel mixture spreads a wall of flame—*flame front*—rapidly, but smoothly, outward through the mixture until it is consumed. But, in some cases, the last part of the the compressed mixture, the *end gas,* can explode before the flame front reaches it.

If the end gas temperature is raised too high, it will explode *before* the flame front reaches it. The resulting shock load, if severe enough, will increase bearing wear, break piston rings or destroy various engine parts.

Why remove tetraethyl lead from gasoline? Because, in part, "lead" emissions may cause cancer. Another reason is that it fouls catalytic converters. The catalytic converter is a primary part used to limit exhaust emissions. Using leaded fuel will eventually render a converter useless.

Performance-car enthusiasts have learned first-hand how fuel octane imposes a practical limit on CR. The high CRs of the late '60s were possible without spark knock because of the 100+ octane fuels available at the pump. Today, a CR over 9:1 may be pushing the limits of 90-octane fuel. Further, the formula for computing octane has changed, and today's 90-octane variety is of lower quality than the 90-octane of 10 years ago. (For historical perspective, CRs were about 7:1 in the 1930s, and as low as 4:1 in the earliest years of the century.)

In practical terms, modern naturally aspirated engines are limited to a CR of about 9:1 because of the lowered quality of readily available fuel. If you are planning to increase your engine's compression ratio above this, make sure you'll have a readily available supply of high-octane gasoline or octane booster.

Increasing CR also requires slight changes in ignition curve and carburetor jetting requirements. As a general rule, raising CR requires a more retarded ignition than normal. Because the temperature of the compressed air/fuel mixture increases faster with higher compression, a retarded spark reduces the chance of detonation. Higher compression engines also scavenge exhaust gases more efficiently, so there is a need to compensate for the intake charge not being diluted by exhaust gas. The mixture can be slightly leaner, especially at idle.

OPTIMUM IGNITION TIMING

The British have a fine method for setting ignition timing. Their technique ignores all numbers and measurements, but instead adjusts the timing under actual operating conditions. As a result, the timing is set for the exact requirements of the engine, taking into consideration its camshaft timing, CR, fuel quality and intake and exhaust systems.

The procedure is simple. After tuning everything else, take the car on the road and drive it at normal speeds at full operating temperature. Suddenly, floor the accelerator pedal and listen closely for knock. If you don't hear any, pull over and slightly *advance* ignition timing at the distributor. Get back in, resume speed, then floor it again, listening for knock. Repeat the process until the engine just begins to knock under full-throttle acceleration. When you reach that setting, *retard* the timing at the distributor ever so slightly—by 1° or 2°—and the job's done.

This approach is elegant for the average tuner, because it tunes the specific car for optimum ignition timing and avoids the need for discussions of before and after TDC, CR or combustion-chamber dynamics. It does require a good ear and a knowledge of which way to turn the distributor to advance timing; the opposite direction retards it. The author's Frazer Nash Targa Florio required this approach because it had no marks for setting ignition timing.

It's easy to determine which way to turn the distributor to advance ignition timing with the engine off, slightly loosen the bolt that locks the distributor in place. Loosen it just enough so you have to exert some torque to get the distributor body to turn. Be sure to return it to its original position with a matchmark; otherwise, the engine may not start.

With the distributor slightly loose, start the engine and let it idle. Rotate the distributor a little and notice whether the engine speeds up or slows down. If its speed increases, you've advanced the ignition timing; if it slows, you've retarded it. Note each direction and its effect.

Why does advancing ignition timing increase engine speed? With an advanced spark, the sparkplug fires and ignites the mixture well before the piston reaches TDC. The burning mixture is further compressed by the rising piston and its temperature is raised even higher by the increased combustion-chamber pressure. More BMEP results and causes the engine to develop more power and idle faster.

With retarded ignition timing, some of the gasoline in the air/fuel mixture doesn't have time to burn before it's forced out during the exhaust cycle. Thus, the engine loses power and runs slower.

In general, a retarded spark reduces hydrocarbons and NOx by keeping combustion temperatures—and BMEP—low. But it can cause driveability problems and overheats the engine and engine compartment by dumping still-burning gases into the exhaust manifold. A retarded spark wastes fuel by allowing it to go out the exhaust port, and a richer initial air/fuel mixture is required.

Setting the optimum ignition timing can be fairly simple for WOT operation. This ignores the fact that ignition timing follows a curve. That is, it starts off at some *static*—initial—setting and advances as engine rpm increases.

Ignition advance can be controlled by a combination of springs and counterweights that are selected to provide (nearly) optimum spark timing at all engine speeds. Or, in modern ignition systems, timing advance can be electronically controlled by microprocessor. A modified engine requires a different advance curve from stock, and you can establish one only by trial and error.

Here are some very general guidelines for changing the shape of a distributor advance curve. Lower-rate—weaker—springs or heavier weights produce an advance curve that increases at lower rpm and reaches maximum ignition advance quicker. Higher-rate—stronger—springs and lighter weights slow the rate of advance and the curve reaches maximum advance at higher rpm.

Changing ignition timing is one of the most straightforward modifications you can make, but integrating a non-stock carburetor with an ignition system can take some ingenuity. The ignition ad-

This show-winning '56 Chevy Nomad features chromed 45 DCOE dual sidedrafts, that work in harmony with a pair of Magnacharger blowers. Talk about radical induction! (Ron Sessions photo).

vance on some cars is modified by a vacuum signal from the carburetor; usually a rubber hose connects the distributor and carburetor.

Typically, a vacuum signal from the carburetor is used to retard the spark when the throttle is wide open. That is, there is a slight vacuum signal below the throttle plate. An engine is most likely to knock at WOT; using a vacuum signal from the carburetor to retard timing, then, is an easy and inexpensive control.

Weber Conversion Concerns—There are several *street-legal* Weber carburetor conversions. These kits are designed to replace a stock carburetor, connect to all stock timing and emission control components, and keep exhaust emissions within legal limits. So don't presume a car with Webers is automatically for off-road use and exempt from federal emission regulations.

Some ignition-related connections cause puzzlement because most performance-application Webers don't have hook-ups for them. You're left with a bunch of wires hanging with no place to connect. Typically, the problem is discovered when you find that the original carburetor has solenoids mounted to it with wires leading to a twisted bundle snaking to the firewall.

Current to these solenoids may be activated by the ignition switch. Your decision is whether or not to keep their function. Insulate connections you choose not to hook up because they may have power when the ignition is on. Discussion of emission-related attachments is found in the section on disassembly and repair of the DFT carburetor, page 71.

While it's not an ignition-related item, many automatic transmissions are also linked to the carburetor. Basically, throttle position is used to cause downshifts below a certain speed. The linkage to control the transmission is another item "left hanging" when bolting on a Weber. Clearly, some adaptation and fabrication may have to be done to cause automatic-transmission downshifts with the Weber installed.

Contact a Weber specialist for information about using a Weber with an automatic transmission. A mechanical linkage poses no severe problem. A vacuum source can be obtained by tapping directly into the intake manifold or installing a spacer plate with a suitable tap between the Weber and its manifold.

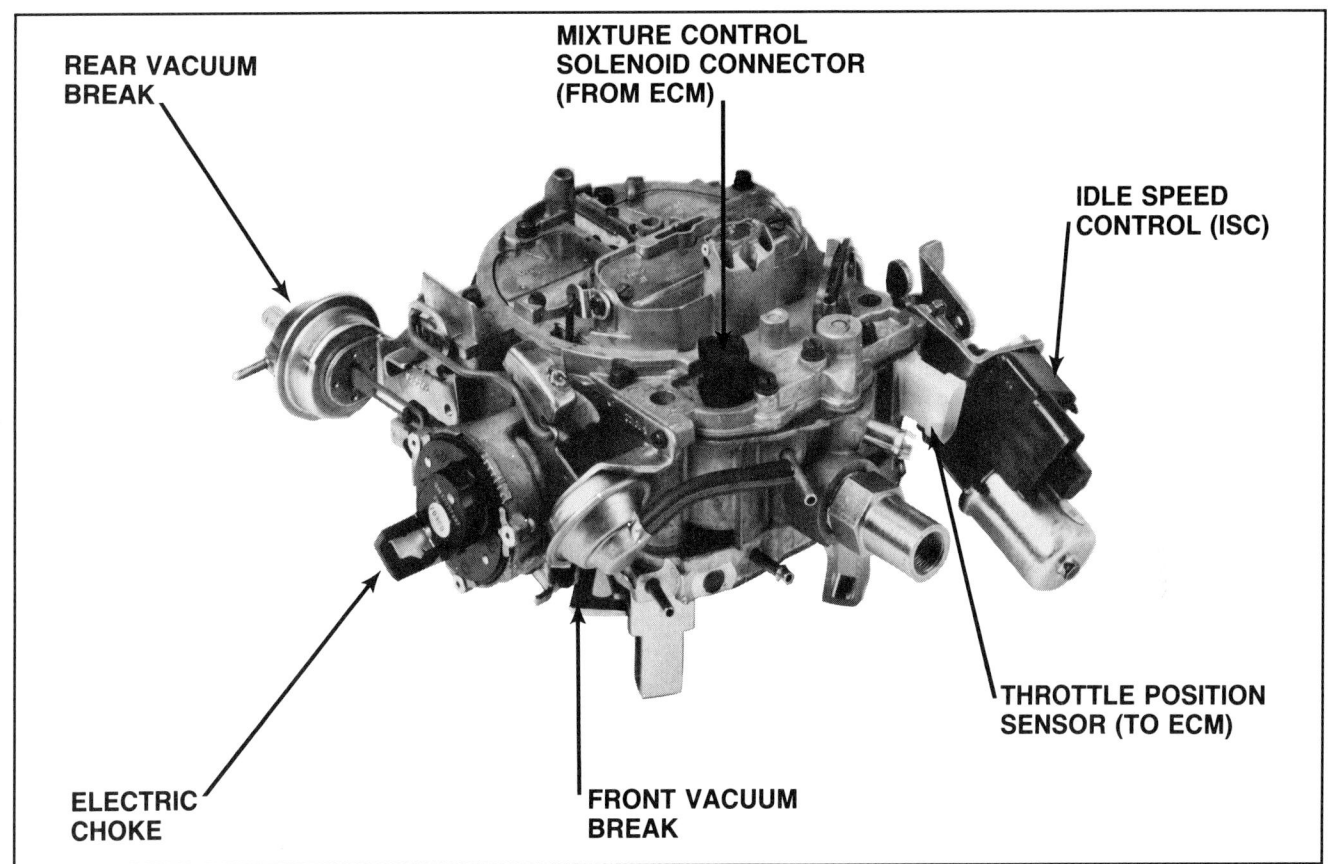

Replacing late-model (post-'81) American carburetor with Weber is difficult. What about all those hoses and wires? Rochester E4ME Quardrajet is linked to Electronic Control Module (ECM).

IMPROVING VOLUMETRIC EFFICIENCY (VE)

Many engine modifications—three-angle valve seats, porting, multi-plane intake manifolds and bundled exhaust systems—are designed to improve the VE. What is VE? If a 2-liter engine could draw in and expel 2 liters of mixture during each intake/exhaust cycle, it would have a VE of 100%. Refer back to Chapter 3, page 18.

Poor breathing is one of the most common characteristics of a low-power engine. Poor breathing means that the cylinder isn't completely filled with a combustible mixture when the intake valve closes. What fills the cylinder, in those cases, is diluted with exhaust gases left over from the previous cycle. These gases dilute the incoming mixture, lower the temperature of combustion, and reduce BMEP. Consequently, improving VE raises BMEP.

Many efficient engines wheeze along with a VE of less than 100%. Getting perfect efficiency from a naturally aspirated passenger-car engine is difficult. Racing engines, on the other hand, can have a VE above 90% at their optimum operating rpm.

Increasing VE isn't simple. It includes work ranging from modifying large dimensions such as increasing the radius of an exhaust-header bend, to very small dimensions such as increasing the angles around a valve seat. The minimum tools required to research VE include a *flow bench* and a method of fabricating and machining prototype parts. So, basic research to improve VE is beyond the capabilities of those of us not in the business of developing engines and their parts.

The question of saving what VE an engine already has is especially critical when it comes to fitting a new carburetor. A Weber has a considerable amount of engineering behind it. But an efficient manifold is necessary to mount it to the engine. Generally, for installing a Weber, you can either modify the existing manifold to accept it, or install a ready-made one that's made for a Weber. TWM Induction offers a comprehensive line of manifolds for many applications; other specialists have them, too.

One inexpensive approach is to buy a spacer that modifies the manifold mounting studs so a Weber will bolt on. But be sure the throttle bores of the Weber are

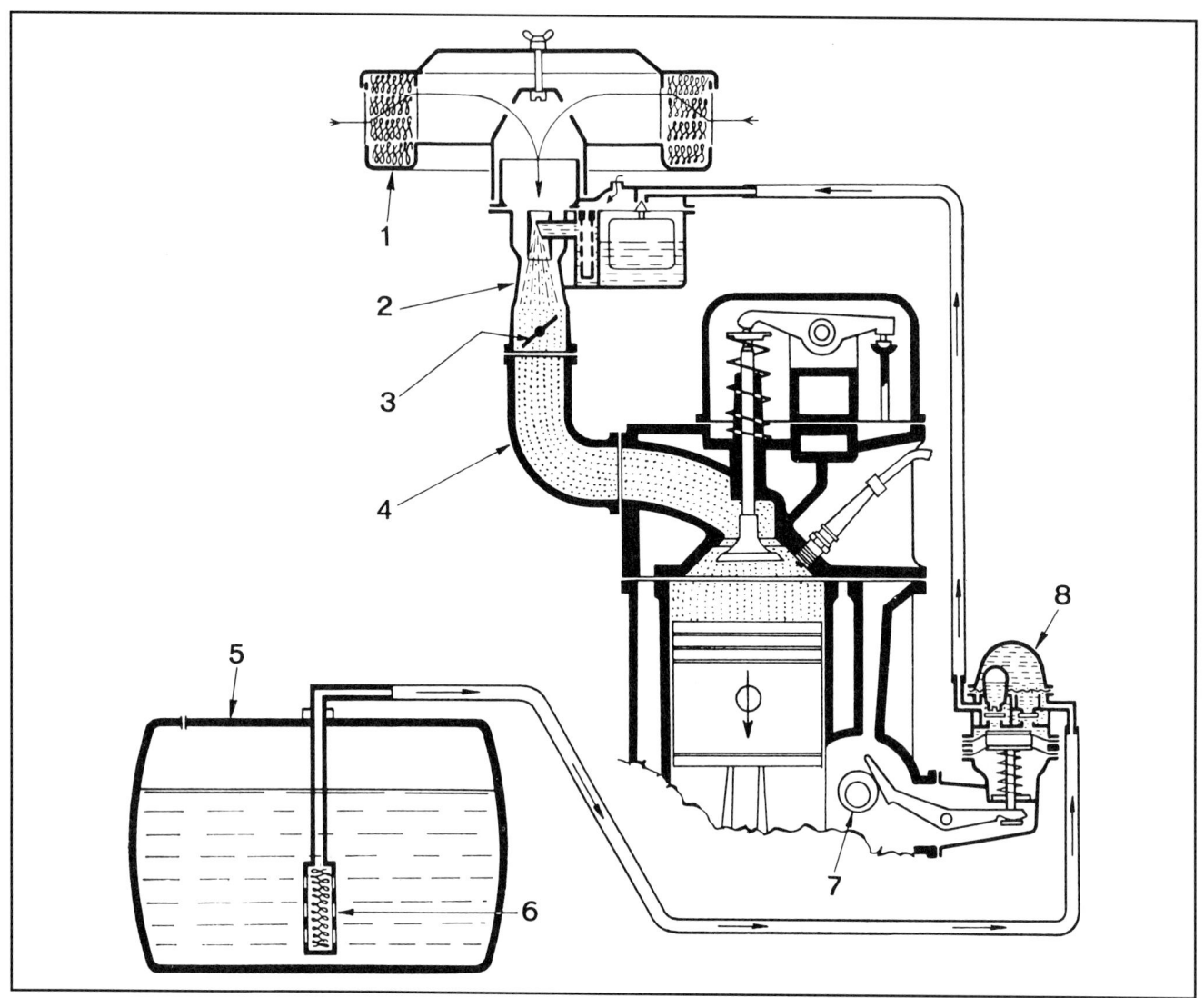

Life in intake manifold is series of pulses, not smooth flow. Ram tuning uses pulses to help fill cylinder. Unsuccessful ram tuning actually keeps mixture from cylinder, reducing performance. Drawing courtesy Weber.

not obstructed by the spacer, and the throttles operate freely. Check hood clearance on downdraft installations—the installation could look great until the air cleaner damages the hood as it is slammed shut the first time.

Ram Tuning—A caution before examining ram tuning: Designing and making a ram manifold probably requires more equipment than you have available. This discussion on ram tuning is more for helping you understand how intake manifolds are designed than to imply that you should try to build one. It should, however, help you choose a manifold to fit your driving style and engine application.

The principle behind ram tuning is to use the *pressure waves* set up by the opening and closing of the valves. Think for a moment of the intake runner with the intake valve closed. There is no airflow, so the condition is static. But when the intake valve begins to open, atmospheric pressure begins to push the air/fuel mixture into the cylinder, as the piston descends in its bore. Airflow past the valve increases. At 90° past TDC, the piston is traveling its fastest, and the pressure differential between that in the cylinder and atmospheric pressure is greatest. At the point of greatest pressure differential, gases in the intake manifold may travel 300 feet per second.

As the piston nears BDC, it slows and the speed of the mixture past the intake valve head begins to slow. Before the mixture actually stops flowing, the intake valve closes. Because the mixture flowing in the intake manifold has momentum, some of it *bounces off* the closed intake valve and reverses direction in the intake runner.

At some point in the intake runner, this pressure wave loses its momentum. But while moving, the wave compresses the incoming mixture slightly, causing a *wavefront*. The intake manifold in a running engine is really populated by wavefronts. Some of those near the closed intake valve are traveling away from the valve, compressing the incoming mixture slightly. The trick is to make the intake runner the right length so that the intake charge reaches the valve head as it opens. In that way, there's a bit of extra air/fuel charge available when the valve opens—it's not bouncing back. This helps fill the cylinder so the engine's VE is increased.

Ram tuning improves VE in a narrow speed range. The ram effect also means that, while there's increased VE at one speed, there's decreased efficiency at other speeds because the waves interfere, rather than help, with cylinder filling.

Most manufacturers that have tried ram tuning—for example, Chrysler in the early '60s—have used it to add mid-range torque. However, depending on the length of the intake runners, power can be made to peak at any speed; in general, the longer the passage length, the lower the speed at which the ram peak occurs.

Matching Port Openings—Generally, you can improve VE by smoothing the intake manifold surface and eliminating any mismatches between the intake-manifold runners and ports in the cylinder head(s). Use a hand-held die grinder to grind away any sharp protrusions in the intake path and ensure the intake passages of the manifold mate perfectly with the ports in the head.

It's surprising how poorly the passages of some intake manifolds mate to the intake ports in the head. To match them, make a template for the head using a sheet of heavy paper. Carefully trim the paper so the mounting studs are perfectly located and the intake passages perfectly match the ports in the head.

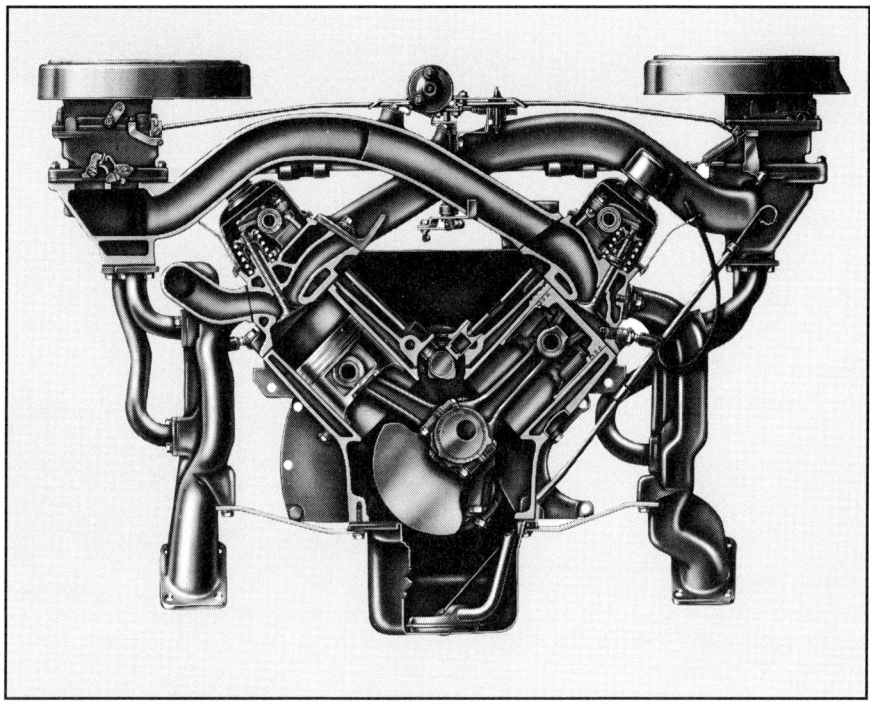

Classic example of ram tuning in a production car: 1963 Chrysler 413-CID 300J Ram Induction engine had eight equal-length intake runners. Each cylinder bank was fed by one carburetor. An equalizing tube connected the two manifold halves. Runner lengths were selected to give 10% torqure improvement at 2800 rpm.

Fit the paper template to the intake manifold, being certain you haven't reversed it. Write MANIFOLD SIDE or HEAD SIDE on the template. Also indicate UP with an arrow. You'll find some of the intake manifold peeking out from under the template. Mark the exposed metal of the intake manifold. You'll be grinding off this metal.

You'll find the paper gasket also blocks some parts of the intake manifold passages. This area is metal to be removed from the head ports. Carefully trim away the template where it blocks the manifold passage. Then refit it to the head, and mark on the metal those sections that require grinding. *Note:* If there's any mismatch, it's always best to have the intake runner smaller than the cylinder head's port.

Grind slowly and carefully, to achieve as perfect a joint as possible between the intake manifold and head. Taper your work smoothly so changes smoothly

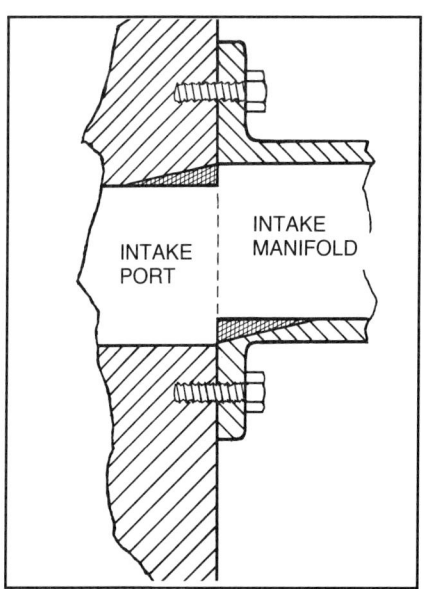

Porting requires matching of runners between intake manifold and intake ports in head. Mismatch causes disruption in airflow, reducing volumetric efficiency. Grinder is used to remove material indicated by crosshatching.

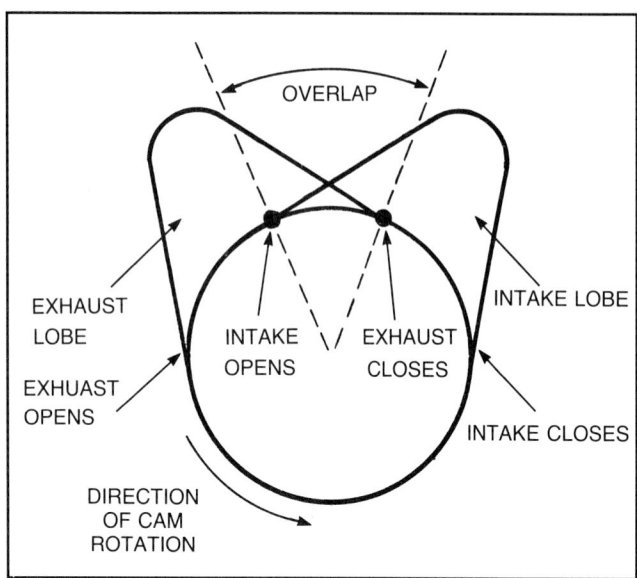

Overlap occurs when both intake and exhaust valves are open. Intake is beginning to open, exhaust almost closed. Pressures in engine keep intake and exhaust gases from going astray during overlap—most of time.

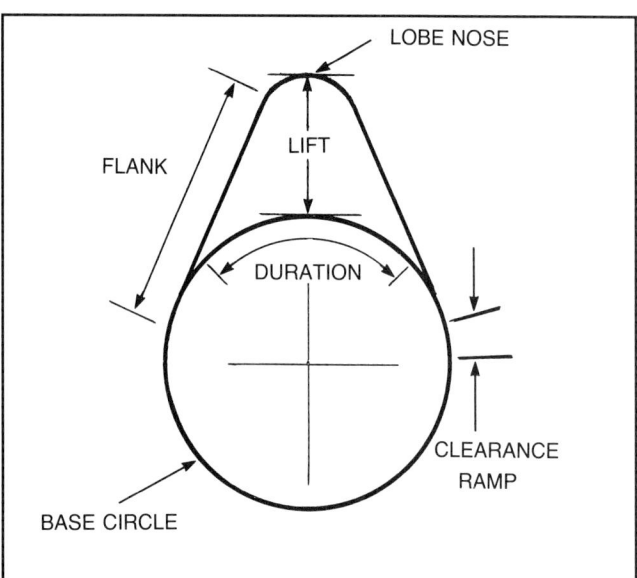

Camshaft terminology: *Duration* represents how long valve is open (allowing for running clearances); *lift* determines how much valve opens.

blend between the joint at the manifold and head. It isn't guaranteed that you can improve anything just by grinding away at all the metal you can reach. Grind as little as necessary to smooth a surface or mating joint, but no more. Grinding until you puncture a water jacket passage is an expensive result from enthusiastic grinding. Bigger is not necessarily better.

Use #400 emery cloth to smooth the surface of as much of the inside of the manifold and head passages as you can reach. Be very careful to clean away any grit after you're all finished. The same technique of matching port openings can be used on carburetor/adapter joints and the exhaust manifold as well.

You'll have to custom-fit the new intake manifold gasket if you've removed quite a bit of material. Precisely match the manifold gasket to the openings, just as you trimmed the paper template.

Camshaft—Next to increased compression and a port/polish job, another popular modification to increase VE is to install a different camshaft.

Practically speaking, a good approach in selecting a cam is to decide exactly what it is you want it to accomplish. "More power" isn't an acceptable answer. Analyze driving conditions and other engine modifications, then talk to suppliers of performance camshafts. They can make worthwhile recommendations if you can supply them with valid data about the engine and your driving application.

Knowing what to ask for gets involved. Cams are usually divided into *street* and *track* categories. In general, it's not a good idea to fit a track cam into a street car. Everyday driveability will suffer. Some cams are referred to by their *lift*. This is the maximum distance they push the valve open. An 11mm (0.433 in.) cam has a rather high lift.

Some camshaft makers refer to cams by their timing specs. A 34-63-63-34 cam begins opening the intake valve 34° before TDC, closes the intake 63° after TDC, begins opening the exhaust valve 63° before BDC, and closes it 34° after TDC. This camshaft has a total of 68° overlap.

Beware of your own expectations. In general terms, every time you add one desirable quality to a camshaft, you have to take away another one. Camshaft design is a compromise between performance at a specific engine-rpm range and driveability. As power is gained from the camshaft, the tractability of the engine decreases.

If you install a cam different from the stock one, you will give up some features the manufacturer designed in. The manufacturer has to satisfy the general public; you don't. Nevertheless, some tradeoffs are necessary. To get better low-end torque, there will probably be less high-end power. For improved high-end power, be prepared to give up some low-end torque. And, in most applications, mileage will be lower than stock.

Changing a camshaft can be the single most dramatic performance modification you can make to an engine short of fitting a supercharger. But just fitting a different cam may not make any difference if the engine can't take advantage of the improved breathing potential. *Note:* It won't supply any more breathing if the carburetor, or any other part of the intake and exhaust systems, is restrictive. Consequently, install a new cam after you've improved the stock intake and exhaust systems.

CAMSHAFT TERMS

A camshaft is essentially a device to open and close the valves in time with the crankshaft. Camshaft design is an exacting science. Opening a valve too quickly runs the risk of losing the "cooperation" of the valve spring. Opening it too much runs the risk of breaking the valve spring or snapping the head off the valve as it slams against its seat. Too soon or too late risks smashing the piston into the valve. Too soon or late can also ruin VE. Just right turns out to be a difficult achievement. Here is an overview of some terms to become familiar with before selecting a camshaft.

Ramp—The ramps are the two flattened parts of the camshaft lobe which are on either side of its peak—*toe.* A camshaft lobe ramp nudges the valve open with increasing speed. At first, only a gentle push is needed to lift the valve against the force of its spring. Once started into motion, the valve can be moved with increasing speed.

The pattern of mixture flow around the valve is determined by the first few degrees of camshaft rotation after the valve begins to open. So the rate at which the valve opens has an effect on total mixture flow and the overall VE of the engine. Finally, the valve must be eased back on its seat to avoid excessive load on the valve head. Once total lift is established, the rate at which tje valve begins to open and close, and the speed with which it moves are both determined by the shapes of the *opening* and *closing* ramps.

Lift—The amount of intake charge passing between the valve and its seat is controlled in part by how far the valve is off its seat. Generally, the more lift the better. The problem comes in getting the high-lift valve fully opened in a high-revving engine. A lot of distance has to be covered in little time. The limitation on lift is the same as on piston stroke—there is only so much available before excess inertia kill the parts.

Duration—The other physical dimension affecting how much intake charge passes between the valve and its seat is the *time* the valve is held open—*duration*. As with lift, you can have too much, both for theoretical and physical reasons. While the valve is open, the piston is also moving. If the valve is open too long, the piston will change direction and start the mixture flowing the wrong way. If duration is way off, the piston may come up and smite the valve with a mighty blow, certainly bending the valve and maybe breaking off the valve head and heavily damaging the engine. So, what's needed is a little bit less than too much duration.

Overlap—Duration is closely associated with *overlap,* which is the time during the exhaust and intake strokes when the exhaust and intake valves are both open. Valve overlap takes advantage of the mixture-flow dynamics of the intake and exhaust manifolds. Changing overlap significantly changes the flow dynamics in both manifolds. There is a point at which overlap is optimum for a specific engine at a specific speed, or when all the pulses work together to improve VE. Conversely, there's a point at which everything "falls apart," and if a camshaft's valve overlap goes in that direction, you're better off with the stock cam.

Increasing valve overlap moves the power curve higher up the rpm range. Decreasing it gives better bottom-end performance. Cams with a lot of overlap produce an engine that must idle at a higher speed, but develops more power at a higher rpm. Excessive overlap can also cause *reversion,* the condition when the intake charge is blown back in reverse and out of the carburetor by the pressure of the exhaust gases during overlap. This reversion is noticeable as a spray of fuel blowing out of the carburetor—*standoff*—under full throttle. It is both wasteful of energy and dangerous as a fire hazard.

Changing a camshaft's *grind*—lobe profile—from stock affects engine performance, especially idle quality and low-end torque. Racing cams can make an engine hard to start, idle rough and have a bad *flat spot* just off-idle so there's a tendency to stumble getting away from a stop. In addition, most race cams give up low-end torque in favor of increased high-rpm horsepower.

Because a performance cam usually increases valve overlap and lift, intake-manifold vacuum is reduced at idle. The engine loses some of its ability to extract fuel from the idle circuit and allows more of the exhaust gases to blow back past the intake valve and dilute the incoming air/fuel mixture. As a result, the engine will run lean at idle unless the idle mixture is enriched by fitting a larger idle jet.

Increased Displacement—If an engine can burn 500 cubic inches of mixture in the same time that it would otherwise burn 350 cubic inches, then it can do more work. There's simply more expanding mixture to convert to mechanical energy.

Increasing either the bore or stroke of an engine to increase its displacement is expensive because it involves precision machine work and assembly time. It is, however, a sure way to increase horsepower. When there's no other choice, and all other modifications have already been made, increased displacement is an alternative.

A vintage Ford Cortina racer competes in SCCA competition with the aid of dual sidedraft 40 DCOE Webers. (Ron Sessions photo).

One of the cheapest methods to increase displacement, incidentally, is to install a larger engine. If you want to increase the size of your powerplant to increase available power instead of supercharging or modifying it extensively, a different engine is a sensible approach. But then you could wind up fabricating and modifying linkages, mounts, fuel lines and wiring to get a Chevy engine running in your MG, for example. The advantage of fitting a larger, stock engine is that it will give more power without ruining driveability—assuming it's not heavier by 10%—and the installation could be simpler. Modifying a smaller engine will get the same peak power output, but with less tractable results.

There are practical design limits to the size of engine's bore and stroke. As bore diameter increases, it becomes increasingly difficult to get the flame front started by the sparkplug to burn completely across the bore. This is particularly true for high-revving engines, because the time available for the mixture to burn is short. As a result, a smaller bore contributes to more complete burning of mixture. But for your situation, maximum bore size is limited by the wall thickness of the bore.

A shorter stroke reduces the speed of the piston as it travels up and down its bore. Consider an engine running at 6000 rpm. The crank is turning 360°, 6000 times a minute, or 100 times a second. The pistons are bobbing once up and once down in the cylinder for each revolution. That means that, in an engine with a 2-in. stroke, the pistons travel at a velocity of 400 inches each second. In an engine with a 4-in. stroke, piston velocity is double, or 800 inches a second—at the same engine speed. As piston velocity increases, so does wear on the piston, bore and piston rings.

A small bore and short stroke is the formula for a very efficient engine, but it's the opposite way to go to increase torque. There is a compromise to be made between efficiencies offered by small bores and short strokes and the outright power of big engines.

One solution is to make an engine with a lot of little pistons moving over short strokes. That's what Ferrari chose with their legendary V12. If you can afford that kind of solution, you also get some lovely engine sounds in the bargain.

The carburetor fitted to a bored and stroked engine will need to flow more air and fuel than the originally installed unit.

Small bores kept ignition demands low on high-speed engine of 1929. Author's supercharged Alfa Romeo displaces 100 cubic inches, develops 100 HP at about 5500 rpm. Bore X stroke of 2.5 X 3.4-in. means high piston speeds. Note blower at front of engine.

You will have to experiment to satisfy the particular applications. In general, the size of both the venturis and main jets can be increased. If more fuel is wanted at higher engine speeds, remember, a smaller main air-correction jet is required to flow more fuel.

Supercharging/Turbocharging—This is a way to increase an engine's VE without changing its physical dimensions. *Supercharging* is a method of pumping more combustible mixture into an engine and making it respond as if it had more displacement. An air pump—supercharger—is mechanically driven by the crankshaft to supply more air to an engine as rpm increases.

Modern materials have allowed the design of a supercharger that is not mechanically driven, but uses the engine's exhaust gases to drive an air-compressing turbine at very high speeds—around 100,000 rpm. This form of supercharging is called *turbocharging*. I review here some of the considerations to keep in mind when turbocharging a engine with Weber carburetion.

The first decision to make is where to place the turbo. Should it be positioned between the carburetor and engine—*draw-through* installation? Or should the carburetor be placed between the turbo and engine—*blow-through* installation? Both setups can be successful if executed correctly.

The primary advantages of the blow-through method are that you don't have to fabricate new intake-manifold ducting, or fabricate new linkage(s) to operate the carburetor. These are major considerations when you get down to the practicalities of an actual installation.

There is also less experimentation required with jetting, because everything in the carburetor must work in a higher-pressure environment. Note that the key word here is *everything*. You must pressurize the float bowl of the Weber as well as the venturi. Otherwise, the high pressure from the turbo in the venturi will actually force gasoline back into the float bowl.

On DCOE, IDF and IDA Webers, the float-bowl breather inlet is a machined surface on the same plane as the venturi inlets. So the only part to fabricate is a simple plenum that covers both the venturi and float-bowl openings. This approach is straightforward and inexpensive because the only construction required is a plenum and an exhaust manifold for mounting the turbo. The passage between the turbo and plenum can be a length of reinforced rubber tube, because the pressures it carries are no higher than those of the cooling system.

Richer mixtures are required with a turbo, in part to keep combustion temperatures low. The Weber is a real asset when it comes to turbocharging because every circuit can be individually adjusted. Larger venturis can be used with turbocharging because mixture velocities are higher. In a blow-through application with boost pressures of about 7 psi, you may find that only a larger main jet will be needed to make everything work fine.

This brief discussion on turbocharging has not touched on intercooling, water injection or ignition timing. If you are considering turbocharging your engine, see HPBooks' *Turbochargers* for a complete discussion of installing a turbocharger with a carburetor. The point here is that turbocharging is a viable way to increase performance, and Weber carbs lend themselves to turbo installations because they are so easy to work with.

Modification Order—There is an order to the way modifications should be made in an engine to get the maximum performance per dollar invested. Many of the modifications that can be done first are also the easiest and least expensive.

In general terms, follow airflow to establish the order in which modifications can be made most efficiently. First, carburetor (don't forget the air filter), intake manifold, head work (porting and polishing) and exhaust system. After that, increase CR, fit a performance cam and increase displacement.

This path is a rough approximation. If the engine is apart for a rebuild, then the order changes slightly. Increasing CR is a significant change you can make because it most directly increases BMEP. But don't do this without considering the octane limitations of the fuel available. Beyond that, there's an old saying: "Two things make an engine go: cubic inches and square dollars."

CARBURETOR SELECTION

Weber publishes a chart of recommended carburetors for specific cars. That material is included in the application charts at the end of this book. In addition, several Weber distributors have created their own compatibility listings, and much of their information is also included in the application charts. If you have a car in the chart, picking the correct Weber is easy.

If it isn't there, find a listing for a car of similar displacement as a starting point. Try to match engine description and packaging—number of cylinders, displacement and configuration, or V-type vs. in-line—as closely as possible. One serious consideration is whether or not you want a downdraft or sidedraft Weber conversion. Dual DCOEs or IDAs are a dramatic and beautiful sight but, for most modern applications, they are not street-legal. On the other hand, you may find that a pair of Weber emission-controlled downdraft carbs such as the DFT may prove legal and still supply the desired performance at a reasonable price.

You may want to choose between DCOEs, IDAs of DFTs. How can you compare what would give similar performance? As a rough guide, use the sum of the venturi sizes when comparing Webers. Just add up the numbers of the venturis themselves. Don't add the *series number* of the carburetor. A 40 DCOE may actually have two 30mm venturis, in which case the total is 60. Similar sums among venturi values will produce similar results on an engine. Although unlikely, a pair of DFTs may be a better choice than a single 45 DCOE, even though the latter looks more dramatic.

Choose more carburetors with smaller venturis when you have a choice. A single 45 DCOE Weber on a 2-liter in-line 4-cylinder engine is not the best application to keep flow rates high. Similarly, if you plan to use your car on the street, consider Webers with a progressive-opening secondary to assure good drivability.

Have a tape measure in hand to ensure that the proposed Weber conversion will actually fit. Even manufacturers have trouble packaging Webers into the engine compartment. On early Alfa Romeos, the Veloce engines with Webers had to be tilted with unique motor mounts to get enough clearance.

If you buy a conversion kit, all this work has been done for you. Otherwise, this discussion is for engineering your own installation. It will, however, help you understand how to evaluate and select the best kit for your purposes.

MANIFOLD SELECTION

First, a warning: Sidedraft Webers such as the DCO series are very sensitive to vibration such as found on four-cylinder engines. Severe engine vibrations cause fuel to foam in the float bowl, which adversely affects float level and mixture strength. For this reason, always use a rubber spacer between the carb and intake manifold to isolate the float bowl. When you do that, you may need to fit a support strut, similarly isolated by rubber, to assist the rubber spacer in supporting the weight of the Weber.

Weber has published a complete set of recommended intake manifold installations for every conceivable type of engine. From the 2-cylinder Fiat 500, through V4 Lancias and American V8s, up to V12 Ferraris and Lamborghinis, these layouts are useful.

At first glance, the illustrations appear painfully self-evident. However, they contain critical information that is easily overlooked. It's important, in studying the illustrations, to note that the forward direction of the engine is indicated by an arrow. And the location of the float bowl, when shown, is always *forward of the venturis*. *Never* place the float bowl of a Weber in any other orientation. Otherwise, the engine will experience severe fuel starvation during acceleration or a hard turn. This caution applies especially if you have an air-cooled Volkswagen engine. Easy conversions with a DGA can place the float bowl to one side, where it shouldn't be.

Note that the illustrations distinguish between simultaneous- and progressive-

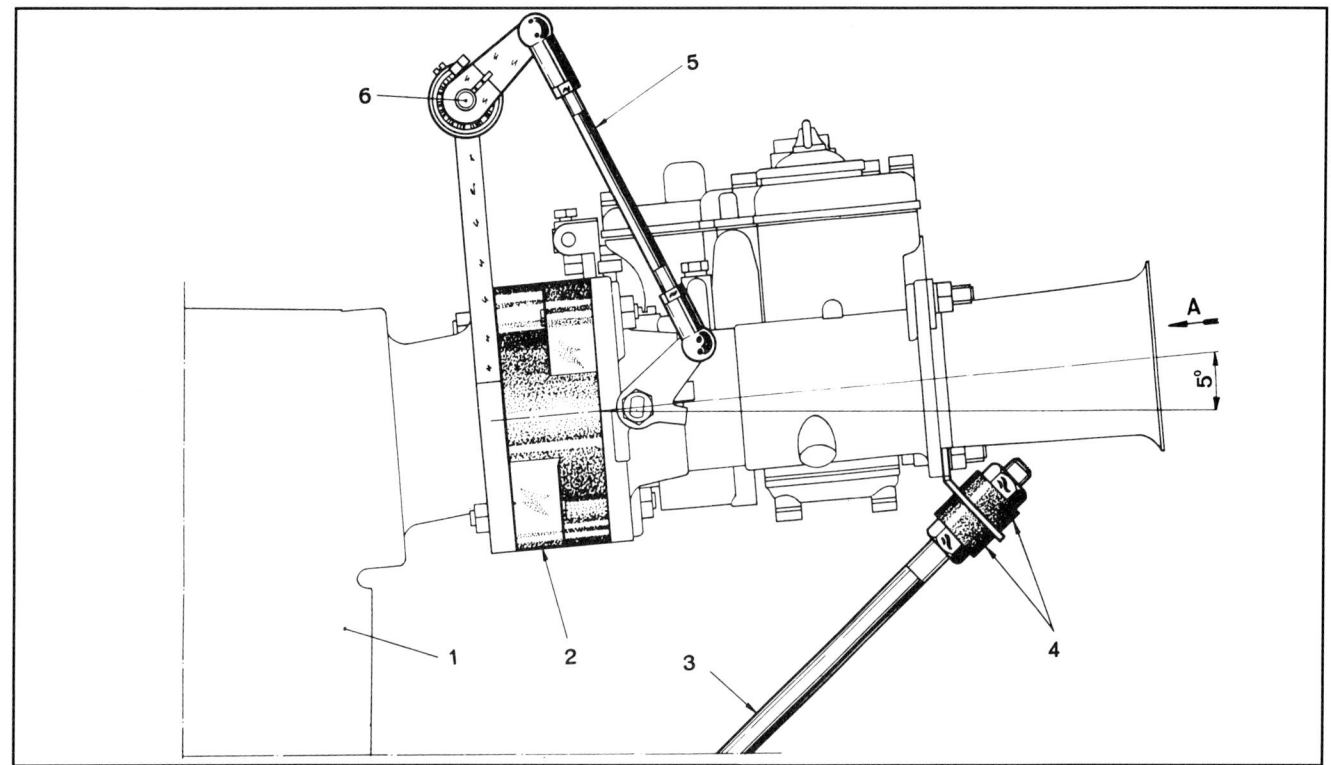

Weber's own illustration of "soft mount" spacer between DCOEs and engine: (1) cylinder head (2) rubber mount (3) support rod mounted to engine (4) rubber isolation dampers, and (5 & 6) linkage. Drawing courtesy Weber.

opening throttle plates. Also note that multiple-plane intake manifolds are indicated for those layouts with simultaneous-opening throttles.

If you can't buy a Weber kit for your car, you may have to make-do with whichever Weber will fit the stock intake manifold. Or, you may have to buy a similar aftermarket manifold and modify it to accept an adapter for a Weber. If you do find an intake manifold that can be adapted, evaluate its design.

Manifold Design—The ideal is that an intake manifold distributes the air/fuel mixture equally to all cylinders. Otherwise, the cylinders nearest the carburetor will run rich and those farthest away will run lean. Because of the need for equal distribution, all intake-manifold runners are approximately equal length and cross-section.

This is clearly a tough challenge when some engines are several feet long and only one or two carburetors are used. Manufacturers typically use baffles or other features to try to equalize flow to all cylinders. When several venturis are used to supply the engine, one or more may be isolated from the most distant cylinders. Those venturis can be jetted richer than those venturis placed nearest the cylinders.

Moreover, the intake manifold is handling a mixture, and fuel atomization is temporary. The gasoline in the mixture can drop out into incombustible pools of fuel along the runners. When the gasoline pools, or *puddles,* the mixture downstream becomes leaner.

To keep mixture velocities high and minimize fuel dropout, designers use cast-in *sumps, ribs* and *dams.* They also heat the intake manifold. This is done with engine coolant or exhaust heat. They also try to keep the runners either horizontal or all equally pointing slightly downward toward the cylinder-head ports. This promotes equal distribution, and drains small droplets into the cylinders before they become large pools.

A small venturi in the carburetor produces higher intake manifold mixture velocities and better distribution. Moreover, the runners in the intake manifold respond to the same principles of flow as the carburetor. Restrictions create higher flow rates and lower pressures, while enlarged cross-sections create lower flow rates and higher pressures.

Just as the carburetor requires air correction to keep the heavier fuel and lighter air flowing at the same velocity, intake manifold design must take into consideration that air turns corners faster than fuel. In sharp turns, the fuel collides against the side of the runner. When it does, it turns into large droplets that form puddles.

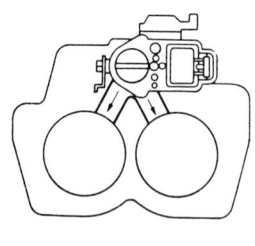
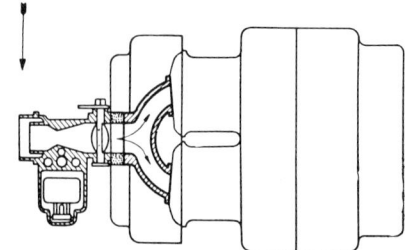
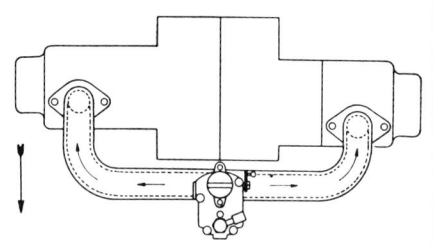

Manifold and carburetor setup for two-cylinder vertical using a single downdraft carburetor. Fiat 500 owners take note: float bowl points toward front of car.

Two-cylinder horizontal layout with sidedraft carb. Subaru owners may be interested. Note that float bowl is foremost.

Weber never did release diagram for air-cooled Volkswagen, but this comes close. Diagram is off twin-cylinder engine, but applies to flat-four as well. Throttle shaft is transverse, parallel to engine, float bowl points forward; anything else is not acceptable.

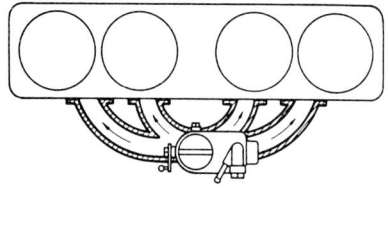
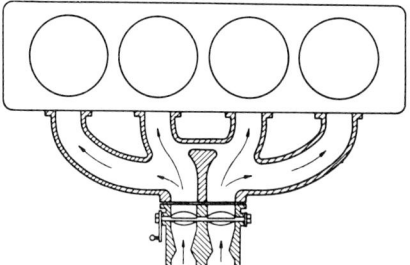
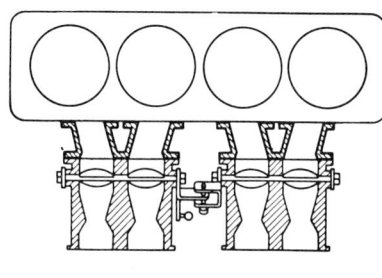

Common installation: single carb on four-cylinder. Throttle shaft parallels engine, float bowl points forward. If throttle shaft were rotated 90°, two cylinders away from idle-jet inlet would run lean during part-throttle operation.

Sidedraft on a four: Nothing special here about float-bowl location, but note important restriction between throats provided by manifold. Plenum manifold, in this case, would permit excess mixing of pressure waves, cause reversion and ruin tuning.

One venturi per cylinder: two DCOEs on four-cylinder engine. Independent-runner design permits maximum tuning possibilities. This is a classic racing configuration for this engine.

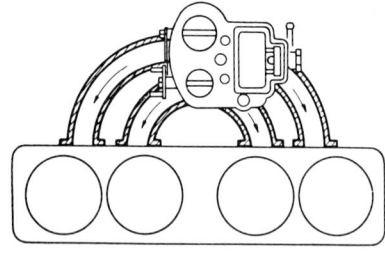
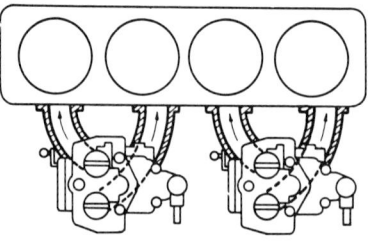

Progressive-opening two-barrel on a four: Primary (small) venturi is closest engine. Float bowl faces forward. This is typical configuration for *street legal* Weber modification, but be sure Weber you buy has needed fittings.

Twin two-throats on four cylinders: Throttles are simultaneous, not progressive, so setup is really downdraft version of independent-runner setup.

Drawings courtesy Weber.

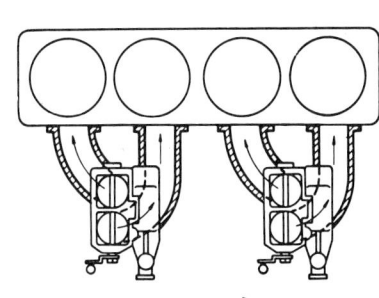

Another variation on two synchronized-throttle carburetors on a four-cylinder engine. Independent-runner design is maintained.

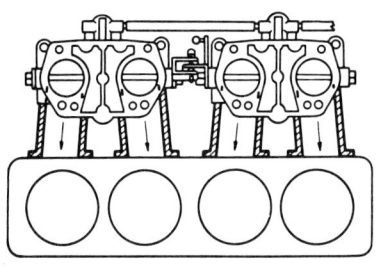

How IDAs fit on fours: independent-runner manifold design in a downdraft, high-performance configuration.

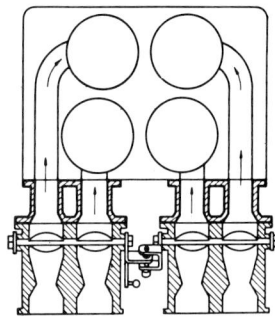

Owners of old Lancias and Ariel square fours; here's how to fit DCOEs. Manifold could not bolt to engine, as shown, but would be independent tubes.

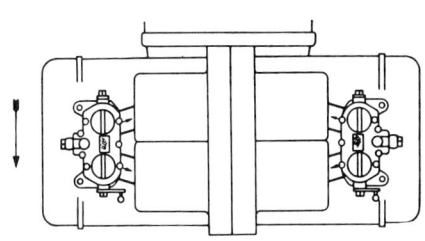

Porsche and modified VW practice: Note orientation of throttle shafts that parallel axis of engine. Floats between venturis make proper setup.

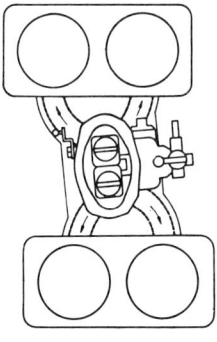

Stock Lancia setup for plenum manifold: Note that throttle shafts match direction of travel, float bowl is forward.

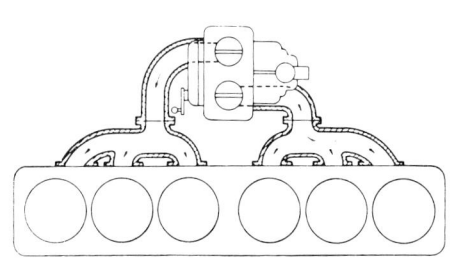

Synchronized throttles, one Weber on a six-cylinder engine: Manifold is actually two separate circuits.

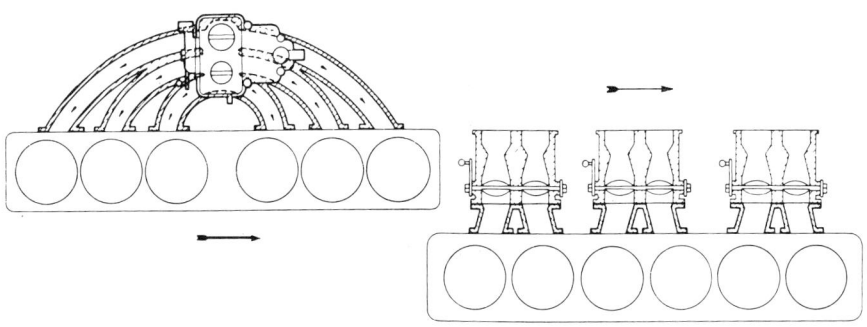

Progressive throttles on straight-six. Note that primary venturi is closer to engine, float bowl faces forward on plenum manifold.

Jaguar, Datsun Z and Stovebolt Six owners' dream: three DCOEs, independent runners on straight-six.

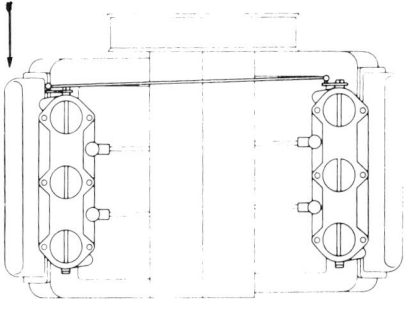

Porsche practice: IDA 3s on flat-six. Corvair owners, take note.

Drawings courtesy Weber.

129

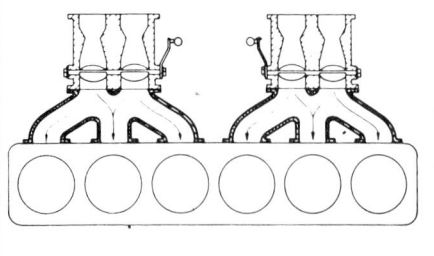

Two DCOEs fit on a six. Not as optimum as three carbs, two still maintain plenty of breathing area.

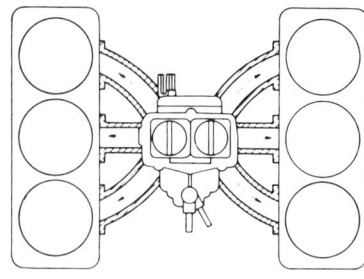

V6 engines get two-throat downdraft Weber with synchronized throttles like this. Plenum chamber permits throttle shafts to be in line with forward axis, float bowl leading.

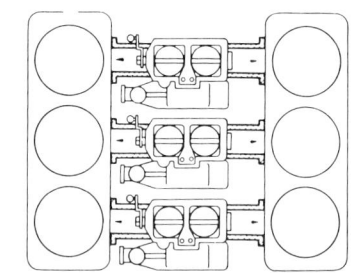

Three DCNs, or similar, on a V6: Independent-runner manifolding and space requirements require transverse throttle shafts.

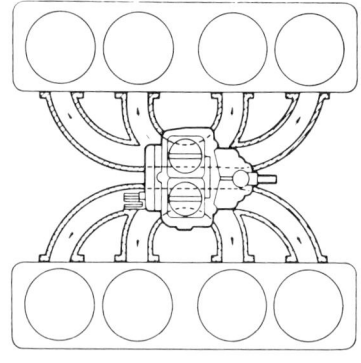

V8 aftermarket conversion places synchronized-throttle downdraft Weber on two-part manifold; not a common practice.

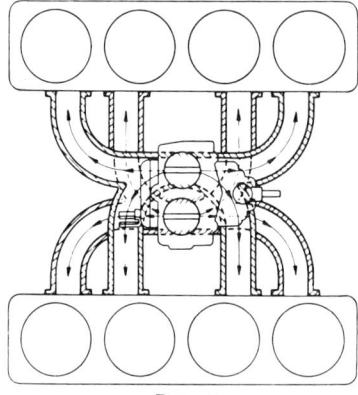

More common V8 practice is dual-plane manifold with synchronized carburetor throttles. This configuration takes better advantage of pulses in manifold.

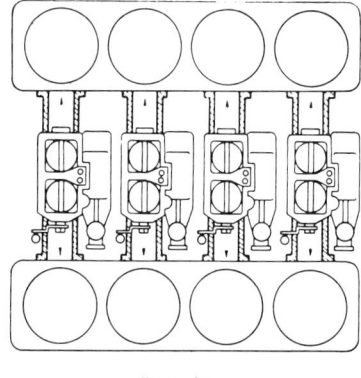

Four twos with synchronized throttles on a V8—classic.

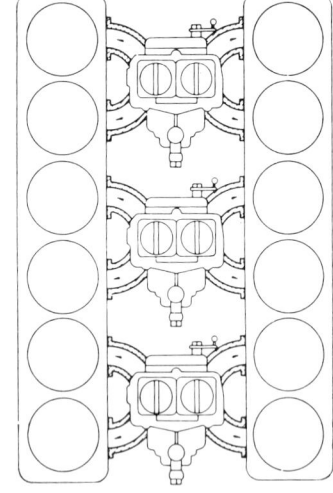

Ferrari practice: three DCN two-barrel carburetors on a V12. Small plenum is beneath carburetor, idle-jet outlet is outboard on each carburetor.

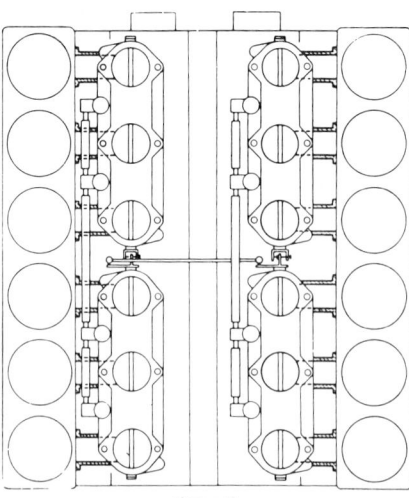

Lamborghini practice: IDA 3s on a V12.

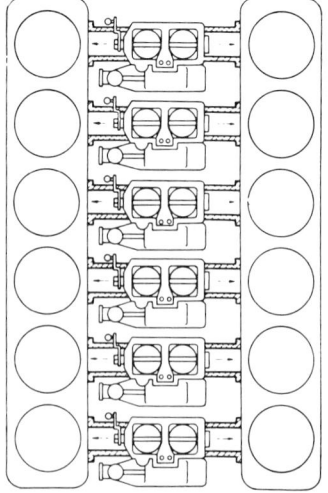

Super-performance setup for Ferraris, old Lincolns: six twos on a V12.

Drawings courtesy Weber.

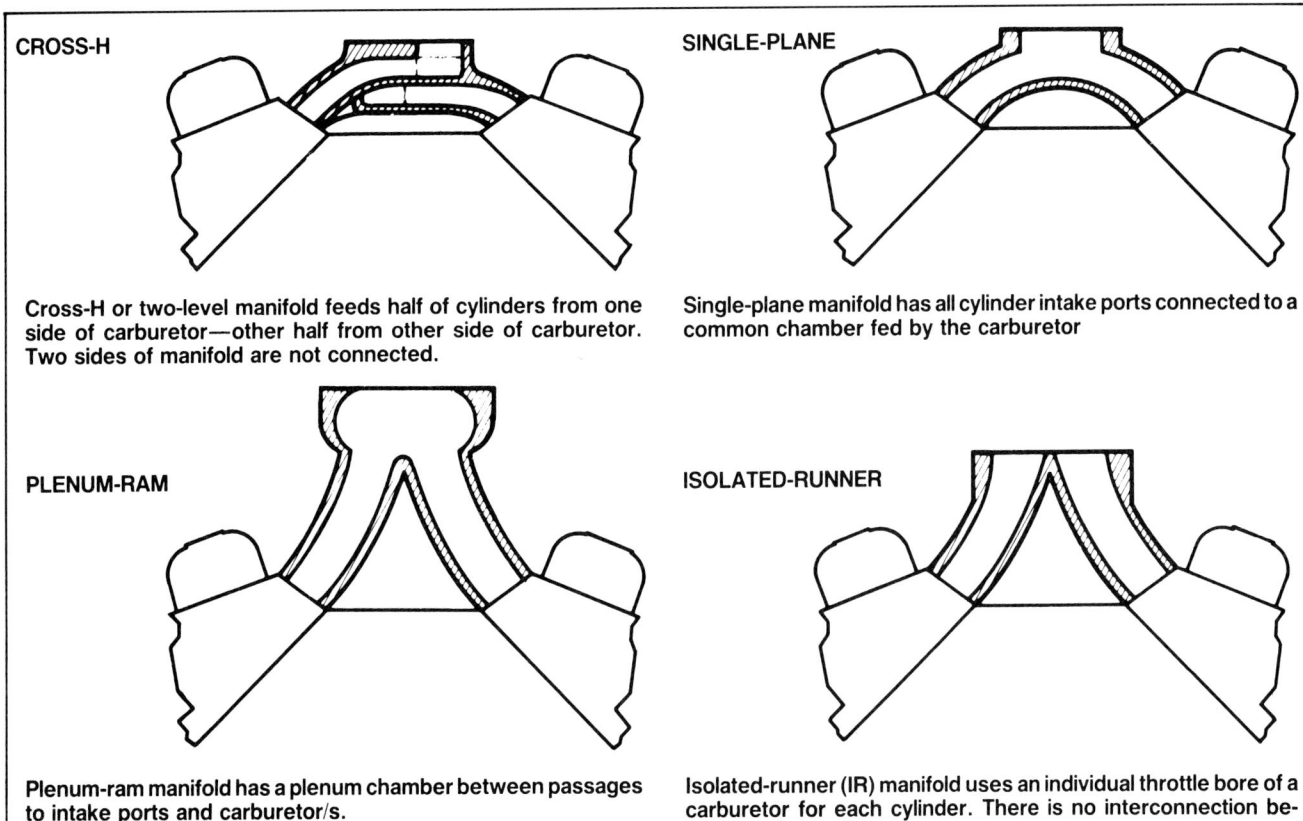

Cross-H or two-level manifold feeds half of cylinders from one side of carburetor—other half from other side of carburetor. Two sides of manifold are not connected.

Single-plane manifold has all cylinder intake ports connected to a common chamber fed by the carburetor

Plenum-ram manifold has a plenum chamber between passages to intake ports and carburetor/s.

Isolated-runner (IR) manifold uses an individual throttle bore of a carburetor for each cylinder. There is no interconnection between the intake ports or throttle bores.

MANIFOLD TYPES

Single-Plane—These are also called *common chamber,* or *runner* manifolds. They have a chamber—plenum—under the carburetor, and individual runners to each port in the cylinder head. The single-plane manifold is simple, but causes problems in multiple-cylinder engines because of the need to supply cylinders in very quick progression.

When adjacent cylinders breathe almost simultaneously, the one that is first on the intake stroke gets the larger share of the fuel. Modern design has reduced the basic problems with the single plane manifold by keeping flow rates high in the runners and taking special care to design runners that promote equal fuel distribution.

Some single-plane manifolds are constructed to take advantage of carburetors with progressive secondary venturis. In those manifolds, a set of small-diameter runners works with the small, primary venturi. A second set of larger runners works with the secondary venturi. In this way, low-speed engine operation is enhanced because the mixture from the primary venturi maintains a high flow rate through the small runners. And, when the secondary comes in, there are larger runners to carry the increased volume.

Dual-Plane—These work with two-throat carburetors with simultaneous-opening throttle plates. One plane of the carburetor typically feeds either the left- or right-hand bank on a V-type engine, while the second plane feeds the other. Thus, there are separate right- and left-hand induction systems fed independently by the two carburetor throats.

The *cross-H* is a common dual-plane manifold for V8s. Its passages are arranged so that the engine breathes alternately from the right or left sides of the manifold, thus spacing fuel distribution needs. Because the mixture mass in a dual-plane manifold is less than in a single-plane manifold, throttle response is crisper and mid-range torque is improved.

Dual-plane manifolds are easily identified by a baffle running along the center of the opening below the carburetor mounting surface. The baffle isolates the two sides of a multiple-venturi carburetor. In specialized racing applications, it can be useful to grind away the top of the baffle. Then, both sides of the carburetor are available to all cylinders for maximum mixture flow. The disadvantage with this modification is that low-end torque is reduced.

High-Rise—These manifolds are designed to take advantage of ram tuning,

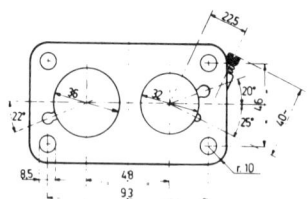
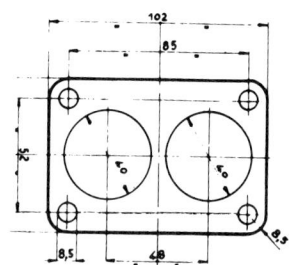

DGV is progressive two-barrel downdraft carburetor for engines from 70—125 CID. DGAS shares same mounting dimensions, has synchronous throttles. Drawing courtesy Redline, Inc.

DFV is identical to DGV, but location of primary/secondary venturis is switched. DFV also has power valve. Drawing courtesy Redline, Inc.

DCNF is synchronous downdraft with interchangeable venturis. Float rides sidesaddle, so orientation of carburetor is important. Drawing courtesy Redline, Inc.

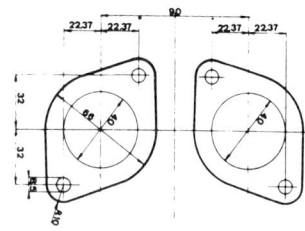
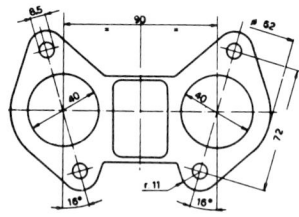

IDF has synchronous throttles, central float bowl for high-performance applications. Drawing courtesy Redline, Inc.

DCOE is classic side-draft carburetor with synchronous throttles, float bowl atop venturi. Vibration-reducing mounting is essential on four-cylinder applications. Drawing courtesy Redline, Inc.

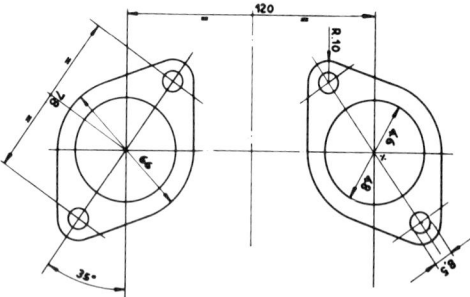
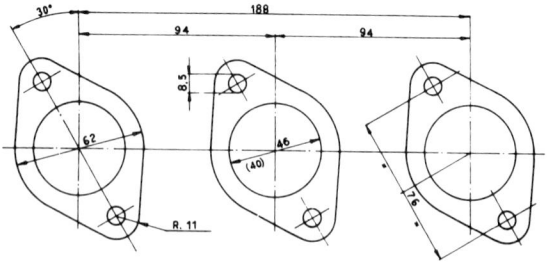

IDA is ultimate synchronous two-throat. No choke circuit, so it's a performance-only application. Drawing courtesy Redline, Inc.

IDA 3C offers one more throat than IDA, but throats are more closely spaced. Drawing courtesy Redline, Inc.

WEBER MOUNTING DIAGRAMS

Attaching a Weber to an aftermarket manifold, especially when the manifold is not made specifically for the Weber, depends on being able to match up manifold's bolt arrangement to Weber's mounting flanges. Diagrams should help you select the easiest Weber to mount. Remember that throttle shaft and float-bowl orientation are critical elements of a properly-designed installation.

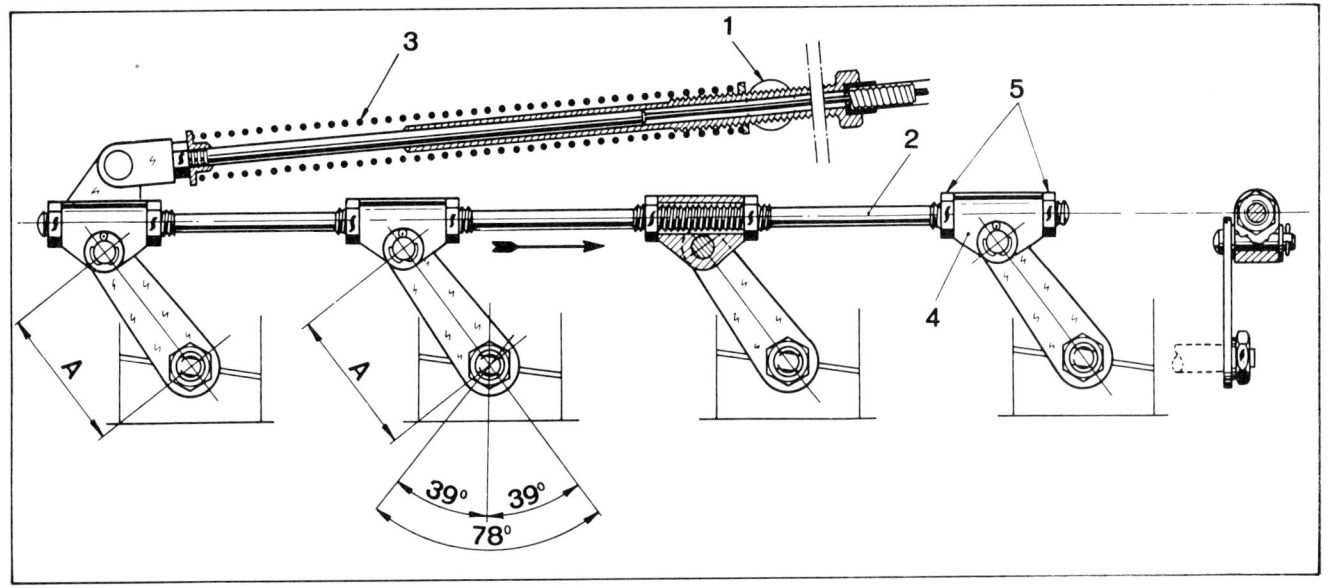

Weber linkage suggestion for Ferrari type four-carb setup, using flexible cable from accelerator pedal: (1) outside of cable, (2) adjustable threaded shaft, (3) accelerator return spring, (4) internally-threaded adjuster block, and (5) locknut. Drawing courtesy Weber.

discussed earlier in this chapter. A true high-rise manifold has equal-length runners that are tuned to help fill the cylinder.

In contrast, the *high-rise* manifold uses a plenum that is 1-1/2—2-in. tall directly under the carburetor. This plenum is used to cancel directional effects the mixture may have picked up in the carburetor. Note that a partially opened throttle plate forces the mixture to flow diagonally across the plate and at an angle into the plenum. The plenum improves distribution in this case.

Isolated Runner—Also called *independent-runner* manifolds, these are for performance applications. They make an exclusive, one-on-one matchup between venturis, runners and cylinders. Such manifolds have an independent intake path for each cylinder.

A carburetor used on an isolated-runner manifold must be larger than for other manifold types because pulses in the runners tend to cause standing waves or reversion. This reduces the flow capacity of the carburetor. A large-diameter venturi overcomes the reversion effect. Velocity stacks on top of the carburetor can be used to contain the fuel—standoff—being pulsed back out.

LINKAGE GEOMETRY

There's one cardinal rule when installing linkages: Don't rely on the bearing surfaces of the Weber's throttle shafts to support much pulling or pushing. The only action the linkage should do is rotate the throttle shaft on its axis. The bearings aren't designed to withstand the force of a cobbled linkage, which tries to pry the carburetor off its mount while operating the throttle.

When bolting on a Weber replacement, position the Weber so its throttle bellcrank works in the same way as the carburetor you removed. Make sure the float bowl faces toward the front of the car. If you can't figure out a way to meet both of these conditions, try placing a bellcrank at the opposite end of the Weber's throttle shaft to replace the original spacer. If you can't do that, use another Weber model, or fabricate a throttle linkage that works.

If you're lucky, a conversion kit's linkage may be adaptable to your installation. If not, contact a Weber specialist for help with linkage construction. An established specialist should have plenty of experience installing Webers in different cars.

Many engine swaps require the greatest engineering when it comes to completing the carburetor linkage. All that figuring about engine mounts and exhaust-manifold clearance is not so hard compared with figuring out how to operate the throttle. Don't scrimp on time or effort to make a workable linkage.

It's essential that the linkage fully open the throttle plates. While this may seem obvious, it's a common cause of poor performance. So, check that the throttle plates fully open when the accelerator pedal is completely down. That way, you are examining the travel of the complete linkage, not just a section forward of the firewall.

Do some basic geometry before selecting and installing a Weber. Begin by determining what part of the existing linkage is usable. Then, measure exactly how much the end of the linkage moves when you floor the accelerator pedal. This is the total available travel of the linkage system. You must make that

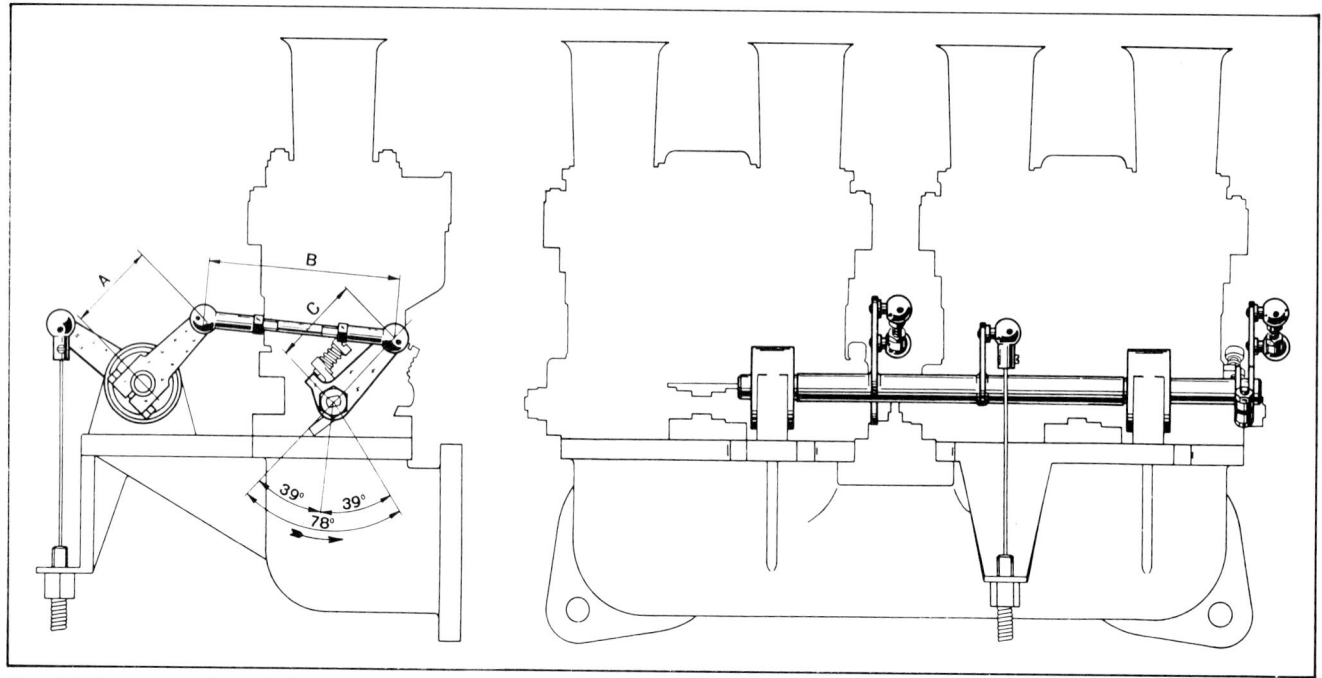

How Weber wants downdrafts set up: Shaft, pillow blocks and linkages are easily fabricated using parts from hardware suppliers. Drawing courtesy Weber.

travel translate to the number of degrees the bellcrank must rotate to fully open the throttle plates.

Some basic geometry and lines on paper are required to determine this. But you can achieve roughly the same result by clamping the Weber on the edge of a table. Use a makeshift spacer underneath that clears the accelerator linkage and keeps the throttle plates from hitting the table when open. Then experiment with a ruler and different length pieces of wire to simulate the throttle linkage.

You can limit accelerator-pedal travel by putting a block under it. The more difficult problem is not having enough pedal travel to fully open the throttle. Shorten the working length of the bellcrank—center of throttle shaft to linkage mounting point—to achieve full-throttle plate opening. The closer the mounting is to the shaft—the smaller the radius—less linear motion is required to open the throttles.

Carefully drill a new hole in the bellcrank closer to its pivot point—the throttle shaft—and fabricate a new connector to join the throttle linkage to the bellcrank.

If a rod-type mechanical linkage is difficult to construct, try using a cable linkage. They are much easier to route.

Again, a major design constraint is not to put excessive load on the throttle-shaft bearings. Test this by fitting the carburetor on its mounting studs, but don't fit any of its mounting nuts. The linkage should fully open the throttle plates without significantly moving the carburetor on its mounting studs.

WEBER SELECTION

Assuming you've selected a suitable manifold and designed the throttle linkage, all that remains is to select the Weber's venturi(s), diameter(s) and jet it correctly. Begin this after taking several test drives. If you've installed a ready-made conversion kit from a Weber specialist, all this tuning homework has been done for you, so you'll have minimal tuning to do.

If, on the other hand, you're interested in peak performance, the final selection of venturis and jets in this type of application is best made after testing. Keep a record of all changes you make in a notebook. This will keep you from duplicating mistakes.

MAIN VENTURIS

As discussed in Chapter 4, Weber has two charts for easily selecting venturis for engines. These charts do not refer to any specific model carburetor. They apply as well to a DGA as an IDA. This information is adequate for most sporting European cars, but hardly scratches the surface for large-displacement American V8 engines.

In these instances, a true dual-plane manifold with twin carburetors will allow you to use the official Weber charts to select venturi sizes for engines up to 5 liters, or about 300 CID. You're only sizing the carburetor for one-half the total displacement of the engine.

If using an independent-runner manifold, which is what most professional Weber conversions use, then jetting is only for a single cylinder. The charts' capacity limits are referenced to a single 2.5-liter cylinder (152 CID). So, you should be able to adapt Webers to a 1215-CID V8. That's probably sufficient for most of us.

Thus, Webers can be adapted to some large engines, provided they are mounted on independent-runner intake manifolds. But the charts have some limitations. Note this modification to them if one carburetor feeds two cylinders that fire 360° apart: Double the actual capacity of the cylinders to determine venturi size. The Engine Capacity chart presumes a maximum of 5000 rpm. The Single Cylinder Capacity chart is for full-race engines. Don't use it for street engines or the car will be over-carbureted.

You can also scan the application tables at the back of the book and select an engine resembling yours in both displacement and configuration. That's a fine place to start to determine venturi and jet size. Err on the side of a smaller venturi, rather than one that's too large. As a general starting point, select a venturi diameter that is 75% of carburetor throttle bore; for example, 40mm bore × 0.75 = 30mm-venturi diameter.

AUXILIARY VENTURI

The auxiliary venturi creates a zone of acceptably high airflow when the mixture flow through the main venturi is too low to provide accurate fuel delivery. If you've selected too large a main venturi, recover some slow-speed driveability by fitting a smaller-diameter auxiliary venturi. In addition, some auxiliary venturis with elongated stacks are available for engines suffering from reversion. Most auxiliary venturis are between 3mm and 5mm diameter. Again, err on the small side when in doubt.

SECONDARY VENTURI

In Webers with progressive-opening secondary throttles and removable venturis, the secondary venturi is typically larger and may be jetted richer. A test for secondary venturi and jet selection is fairly hair-raising, because it should be performed at near-maximum speeds. A flat, deserted road or track is the place to test secondary venturi operation. Another rule of thumb: If in doubt, jet the secondary the same as the primary.

In many instances, it's best to stick with the equipment that came stock on the Weber until you've really sorted out the lower-end performance

SYNCHRONIZING MULTIPLE CARBURETORS

Many mechanics shiver when thinking about synchronizing two Webers. Mere mortals leave the four on V8s or the six on Ferrari V12s to those with divine inspiration. However, synchronization is a straightforward task if you have a little patience.

Actually, there is a chicken/egg problem here. You should have correctly jetted carburetors to synchronize them. But you can't determine the correct jetting unless the carburetors are synchronized in the first place. The procedure is to make one pass at synchronization with the engine off. Then jet the carburetors, and finally go back and re-synch them using only the idle-mixture richness and idle-speed adjusting screws.

Linkage—Begin by setting all independent mechanical linkages to the same measurements. If two rods are used to operate two bellcranks, then make sure the rods are equal length. Use a vernier caliper and be as accurate as possible. Disconnect the links at the throttle-shaft bellcranks.

Idle-Speed Screws—Back off each idle-speed adjustment screw so there is clearance between the end of the screw and its stop on the throttle-shaft bellcrank. Use a 0.010-in. feeler gage to set each idle-adjustment screw exactly 0.010 in. off the stop.

Actually, any dimension is acceptable so long as each idle-speed adjustment screw is set the same. The carburetors are now *zeroed* so all throttle plates are synchronized (closed) and the idle-speed adjusting screws are all equidistant from their stops.

From here on, count the turns you make to the idle-speed adjustment screws

---

> **COMMON-AXIS CARBURETORS**
> Weber uses a spring-loaded link for connecting two twin-throat carburetors, such as the DCOE. There is a single adjusting screw that captures a tab against a spring. The *female* joint (adjusting screw and spring) is attached to one carb and the male joint (tab) is attached to the other carb.
>
> The correct adjustment for this link is to turn the adjusting screw so it just forces the tab against the spring. Turn the screw in and out until you're sure it only eliminates lost motion in the link. Then, use a UniSyn or Synchrometer to check airflow through each carburetor. Both should ingest equal amounts of air. Turn in the adjusting screw on the link—against the spring—to equalize flow between the carburetors.

---

with extreme precision. Note flat-by-flat, if the screws have hex heads, or 45° increments if you have only a screwdriver slot to work with. Turn in each idle-adjustment screw so it just begins to move the bellcrank. Double-check that each adjustment screw was turned the same amount to contact the bellcrank stop. If there's any difference in number of turns, find out why. It's OK to bend to the bellcrank stop a little if you're very careful. Double-check stop clearance by popping the throttles closed. There should be no clearance between any adjustment screw and its stop.

Attach Linkage to Bellcrank—Connect the mechanical linkages to the throttle bellcranks. Check for clearance between the idle-speed adjusting screws and their stops. If there is any, figure out why and fix the problem.

Open Throttle Valves—Turn in each idle-speed adjusting screw 360° so all the throttle valves open very slightly.

Idle Mixture—Count the turns it takes to lightly seat each idle-mixture richness screw. Take the average number of turns and back out each screw from its lightly seated position that number of turns.

Expect on early 40 DCOE 2 and 40 DCOE 18 units to back out the screws about 3/4-turn. Later DCOEs, such as the 40 DCOE 152, require 2-1/2 turns out.

If you have access to an HC/CO meter, adjust each idle-mixture screw by turning it 1/2 turn either way and noting the emissions levels. Adjust idle mixtures until the levels are similar for all screw adjustments; record the number of turns.

Mechanically Correct—The carburetor should now be perfectly synchronized mechanically. The engine may even run. Start it and find out. From now on, and I mean forever more, *NEVER alter one adjusting screw* unless you make the identical adjustment to each of the others. This applies to both idle speed and idle richness adjustments.

FLOAT-LEVEL ADJUSTMENT

Before beginning to jet a Weber, verify that float level is set correctly. Float level controls the depth of the fuel in the float bowl and well and, therefore, the point at which the main circuit starts. A low fuel level restricts main-circuit startup; too high a level can overrun the well and dump raw fuel into the throttle bore.

Weber typically specifies two checks on float setting: the point at which the float is nearest the float bowl cover, and the farthest point it can travel from the cover. The measurement on DCOE carburetors is taken with the float bowl cover gasket in place.

JETTING

Main Jet—Review the discussion in Chapter 4 about jet selection. To begin, establish a primary venturi main jet size. Use conversion kit listings as a guide. Or, select a preliminary main jet according to this rule of thumb: Choose a main jet four times the value of the its venturi diameter; for example, a 30mm venturi equates to a 120 main jet.

Pay no attention to top end. You're concerned with mid-range power and how the engine pulls from a standing start. Main-jet size is the key to everything else, so select it with care and lots of experimentation. Frequently check tailpipe deposits and read the sparkplugs from time to time to make sure you're not cooking—or fouling—anything.

Once you've established the initial main jet size, get a set of the next-size main jets, both smaller and larger. Choose jets in increments of five. If you started with a 120, buy 115, 125, 130 and so forth. Try them. If driveability improves, great. If not, it was worth the effort to find out.

Air-Correction Jet—Don't change emulsion tubes unless you have access to a chassis dynamometer and an HC/CO meter. For your needs, the air-correction jets control top-end mixture strength. Remember, the *larger* the jet number, the *leaner* the top-end mixture. That's just the opposite direction of the numbers on a main jet. Larger air-correction jets mean more air; larger main jets mean more fuel.

Start with an air-correction jet with a value of 60 larger than the main jet. For example, if using a 120 main, install a 180 air-correction jet. Go for a short, but fairly high-speed drive, paying attention to throttle response in the range of 3000—5000 rpm. Note the condition of the tailpipe and sparkplug deposits. Install a smaller air-correction jet for a richer mixture, or a larger one for a leaner mixture.

Idle Jet—Now for the idle jet: You're not adjusting the idle-mixture screw, but changing the jet that feeds both the idle and progression circuits. If you have a severe problem with this circuit, you'll know it the moment you try to drive away. The engine will cough and stumble until it reaches about 2000 rpm. At the same time, it will also *surge*—which is a sudden burst of power that goes away just as suddenly.

Checking the idle jet is detailed in Chapter 4, page 39.

Secondary Venturi Jet—If using synchronized-throttle carburetors such as the DCO or IDA, jetting is the same for each venturi.

Jetting the secondary venturi in a progressive-throttle carburetor, such as the DFT, follows roughly the same procedure as for the idle jet. There is an additional aid, however, for knowing when the secondary starts. You may feel a slight increase in resistance when pressing the accelerator pedal far enough to open the secondary. Look for a smooth transition when the secondary opens— no surges or flat spots.

The main jet for the secondary is typically a bit larger than for the primary. Its air-correction jet is frequently the same size as the primary air-correction jet. If experimenting with the secondary jets, continuously check the color of the tailpipe deposits. Try for a chocolate-malt brown color.

SECOND SYNCHRONIZATION

Idle Adjustment—Now, make the final idle adjustments, presuming the engine is properly jetted. First, *equally* adjust all the idle speed screws to give as slow an idle as possible. Then, *equally* adjust all idle-mixture screws to get an idle as fast as possible.

Don't turn one screw just a quarter-turn more or less than the others. **Always** turn each exactly the same. Repeat the entire process three or four times until satisfied you have the best settings possible. If working with an IDA or another carburetor with bypass air screws, refer to the last part of the IDA overhaul section for instructions on adjusting them.

Once you're satisfied you've obtained the best settings, step back and take a solemn oath not to make further adjustments. That's how Webers stay synchronized. See the Alfa and V8 installations in Chapter 7 for more information on synchronizing multiple-carburetor applications.

7
Typical Weber Conversions

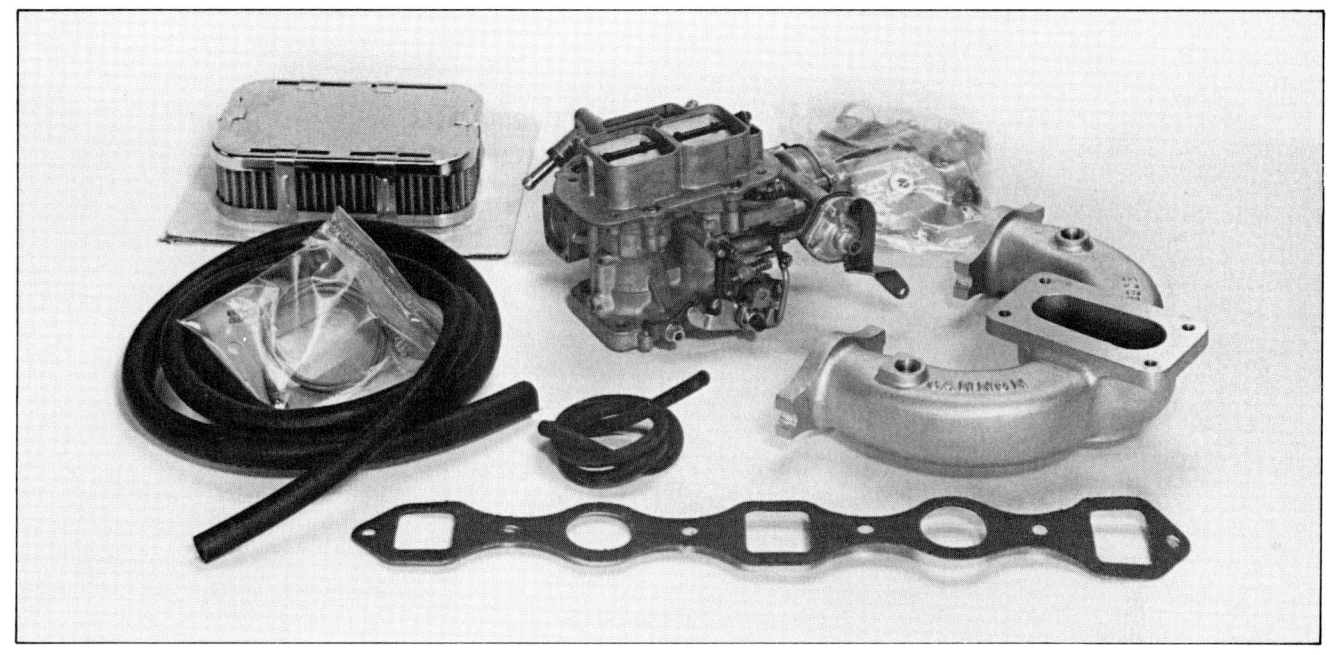

Redline, Inc. DGAV street-legal conversion (pre-'74) for MGB: It passes California emissions. Always lay pieces out and get familiar with them. Small pieces are bagged to avoid loss of parts.

DGAV ON AN MGB

This installation has much to recommend it. First, it puts a Weber on a car that significantly benefits from the conversion. Second, it represents more of a challenge than simply bolting on a replacement carburetor and illustrates how a properly engineered installation should go together. The installation produces a street-legal car that will pass even California's strict pollution-control laws (pre-1974). Finally, it is the carburetor picked for representative disassembly in Chapter 5.

The DGAV and its manifold are part of a kit marketed by Redline. Everything for the conversion is included in a single box; it includes a clearly written instruction sheet. The instructions cover MG models from 1968—74. While the MG engine didn't change much during that time, the emission control laws became more stringent. So, the kit has to span a variety of possible configurations and occasionally you must improvise to make the pieces fit. This particular MG engine had twin SUs, which were replaced by the single downdraft DGAV.

Only hand tools are required for the conversion. And as more evidence the kit is well-engineered, there's no cutting or welding required. The complete transformation can be done in about three hours.

The instructions build confidence, but there can be surprises that aren't explained, depending on the installation. In this particular car, I found that the exhaust manifold could not be removed from the engine without also unbolting it from the exhaust pipes. Though not mentioned in the text, the kit did contain two new exhaust flange gaskets, proof that the folks at Redline really pay attention to actual installation needs.

Begin the installation by laying out all the pieces and making sure they're all there. Read the instruction manual through, mentally assembling the parts. Begin work in a clean, well-lighted place and take your time.

Remove four screws and stock cleaners come off as an assembly.

Accelerator-cable trunion is attached to throttle linkage with cotter pin: Remove with needle-nose pliers.

REMOVING STOCK INDUCTION

Preparation—First, for safety, remove the gas cap and disconnect the battery. Also, do all the work on a cold engine. This ensures that 1) you don't burn your hands and 2) no hot parts warp when removed. Clean the engine before starting. A clean engine can make it easier to find fasteners that were once buried under grease and ooze. Drain the radiator. Reassembly would be a fine time to change the coolant.

Air Cleaners—Remove the stock air cleaners by unscrewing the four bolts holding them against the SU carburetors. The bolts on the front carburetor air cleaner screw into a yoke that also holds the manual choke cable bracket. When the bolts are withdrawn, the bracket will fall free. On the rear carburetor, hold the yoke with one hand to steady it while unscrewing the final air cleaner bolt.

Next, remove the vent hoses to the crankcase and charcoal canister, if fitted.

Linkages & Cables—The DGAV has an electric choke; remove the stock choke cable. Or, simply tape it out of the way, against the inner fender panels.

Remove the throttle cable from the linkage by pulling the cotter pin from the trunion that clamps the cable. When the trunion is free, the accelerator return spring will also snap free.

Remove the fuel line from the carburetors at its rear fitting and plug it with a dowel to keep fuel from spilling.

Pull off the distributor vacuum line to the carburetor manifold and plug it.

Remove SUs & Manifolds—Unscrew the four 13mm nuts holding the carburetors to the intake manifold. Grasp the heat shield and give it enough of a tug to free the shield, carburetors and their spacers from the engine.

Remove the assembly as a unit by holding both the carburetors and heat shield and pulling everything off the studs at once. The connecting links for both the throttle and choke shafts will probably fall free. Pick them up from underneath the car.

Remove the *diverter valve* and all its hoses as one unit. Sometimes called a *gulp valve*, this emissions-control piece prevents backfires during high-speed deceleration. Save it.

The intake and exhaust manifolds are held on by six nuts. Lubricate the stud threads with penetrating oil before removing the nuts to ease removal and avoid breaking a stud. Keep track of the order in which you remove the nut, washer and larger clamping washer.

Pull free the intake manifold. Be careful not to let dirt fall into the two intake ports. Stuff clean rags in the ports if the engine will sit for some time before the carburetor and manifold are replaced.

The exhaust manifold is removed because the intake and exhaust manifold gasket is one piece. The intake manifold gasket should be replaced when the new manifold is bolted on. Consequently, the exhaust manifold must be removed so the entire gasket can be replaced.

The exhaust manifold may foul against the steering shaft so that you can't just pull it off its studs. Remove the six flange bolts attaching the manifold to the exhaust pipe and then remove the manifold. Remove the old manifold gasket, scrape the surface so that it is completely clean and then put a new gasket in place. Refit the exhaust manifold to the engine, holding it in place with nuts at either end.

Four easily accessible 13mm nuts secure both carburetors to stub manifold.

Heat shield and carb spacers come off as a unit if you hold as shown.

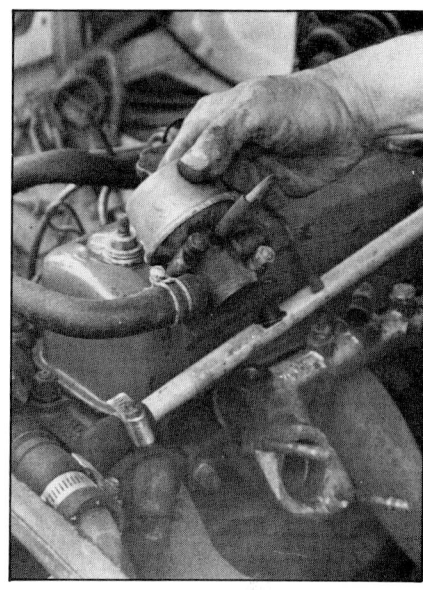
Gulp valve—diverter valve—is used by conversion, so remove it carefully.

Nuts that attach stub manifold to engine may be rusty. Soak with oil before trying to remove.

Intake manifold removed, exhaust manifold slips off studs. Removal of exhaust pipe at manifold flange is necessary. Clean gasketed surfaces with wire brush.

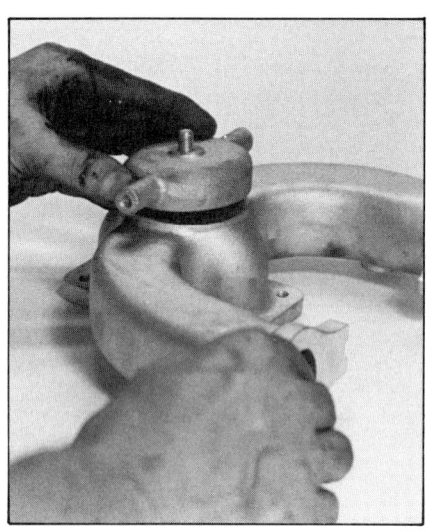
Manifold heater is installed with rubber O-ring.

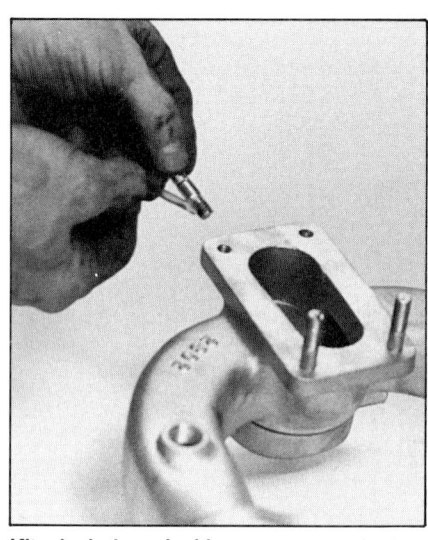
Kit includes locking compound—just enough. Use it for locking manifold studs.

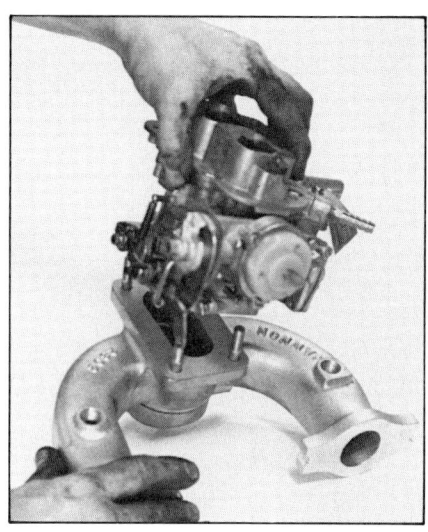
Slip gasket, carburetor on manifold as shown. Plastic choke heater points toward head.

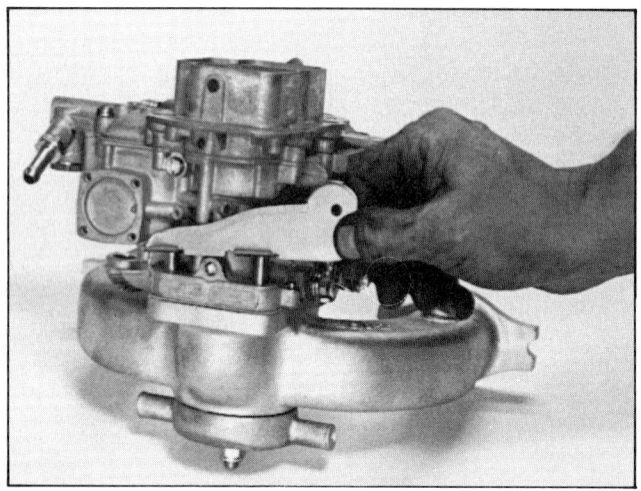
Throttle-cable bracket fits on outside of carburetor, slips on studs.

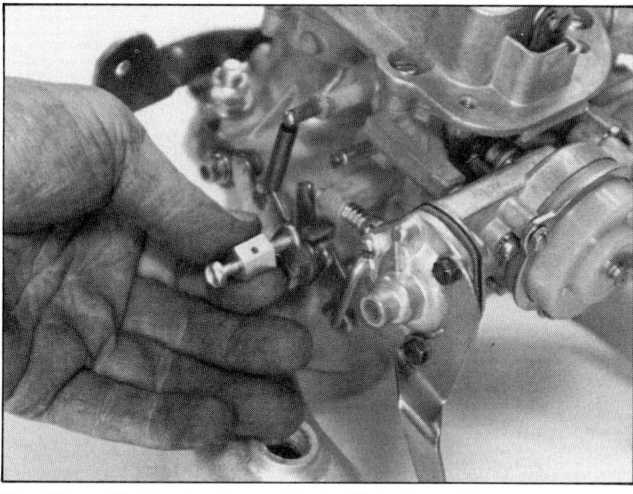
Throttle-cable trunion is attached with locknut. Be sure trunion is free to rotate.

ASSEMBLE DGAV & MANIFOLD
Manifold Heater—This kit features a water-heated intake manifold—another concession to lowered emissions. The manifold heater is plumbed into the heater hoses and is attached to the manifold with a single stud, aluminum washer and Nylock nut. It is sealed with a rubber O-ring. Position the inlet and outlet nipples so they will point fore/aft when the intake manifold is installed on the car.
Mounting Studs—Install the four carburetor mounting studs using a drop of the locking compound supplied in the kit. Coat the stud with locking compound and then screw the stud home into the manifold. Don't use all the compound on this step. You'll need it later on.
Mount Carburetor—Slip the carburetor gasket onto the manifold, then install the carburetor. Point the choke assembly toward the mounting flanges of the manifold.
Throttle Cable—Slip the throttle cable bracket onto the two *outboard* studs and secure the carburetor with four spring washers and nuts. Don't overtighten the carburetor nuts. Just screw them down until the spring washers are fully compressed and snug.

Manifold is tightened down. Large brass fitting for diverter-valve plumbing goes on with locking compound.

Note smooth bend in hose between diverter valve and manifold. Valve must be positioned to avoid kinks.

Add vacuum-delay valve, black side to front of car.

Attach the throttle cable trunion to the bellcrank with a Nylock nut. When tightening the nut, leave the trunion free to rotate as the throttle cable moves through its arc. Tighten the nut so the trunion doesn't wobble, but no tighter. Thread a machine screw into the end of the trunion.

Slip one end of the throttle return spring (supplied in the kit) into the end of the bracket that extends from the choke housing. The other end of the spring slips over the throttle cable trunion.

INSTALL DGAV & MANIFOLD

Prepare Manifold Flange—Put a small bead of silicone gasket sealant around the intake manifold flange. Remove the rags from the intake ports and slip the new manifold/carburetor assembly in place. Reattach both the intake and exhaust manifolds.

Install Fittings—Two brass fittings are installed next, using locking compound. One has a large nipple, and the other a small nipple. Put locking compound on the large-nipple fitting and screw it snugly into the front threaded opening of the intake manifold. Follow the same procedure for the small-nipple fitting and put it in the rear threaded intake manifold opening.

Diverter Valve—The large diameter hose between the diverter valve and manifold always carries manifold vacuum. Therefore, position it so it isn't kinked and won't collapse. Attach a new hose to the front manifold fitting (larger one), and route it so it makes a 90° bend forward without kinking.

Decide where to mount the diverter valve along the hose route. Directly in front of the rocker cover is a good location. Cut the hose to length, attach the diverter valve, and then attach another length of hose between the valve and the air pump. Secure all attachments with hose clamps.

Install the small vacuum hose between the diverter valve and the rear brass fitting on the manifold. At some point in-line, splice in the vacuum delay valve. Be sure the *black* side of the valve is toward the diverter valve.

Manifold Heater—Now, plumb the manifold heater. Using the new hose supplied in the kit, cut a length to run from the heater nipple on the driver's side of the unit to the back of the manifold. Fit another piece of hose from the manifold

Plumbing coolant to intake manifold requires juggling hose positions for fit. Metal section is used where line passes too close to exhaust manifold.

to the nipple that is part of the lower radiator hose assembly. Secure all attachments with hose clamps.

Charcoal Canister—If the car has a charcoal canister, the canister vacuum signal hose is attached to the hose from the rear brass fitting using the T-adapter supplied. Alternately, the T adapter can be used as a convenient vacuum source for troubleshooting, using a vacuum gauge.

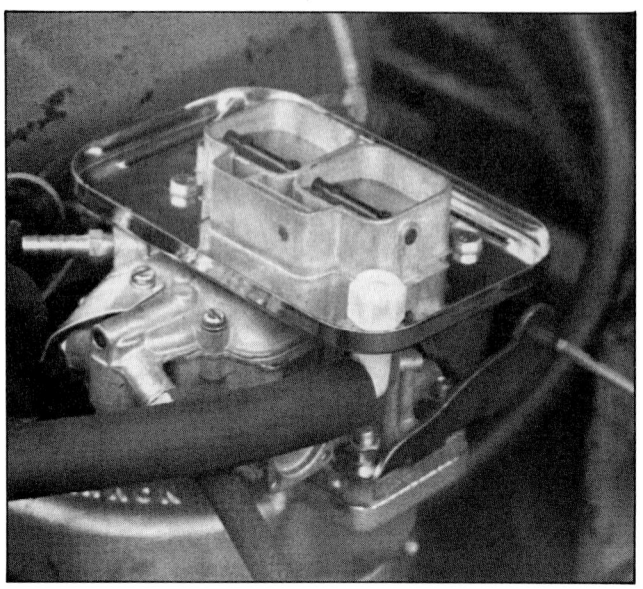
Crankcase-breather hose attaches to white plastic nipple on air cleaner.

Completed installation is neat, sure to pass emissions.

Throttle Cable—Slip the throttle cable into its mounting bracket and attach the inner cable to the trunion on the throttle bellcrank. Route the throttle cable so it doesn't make any sharp turns. It's quite long for this application, so position it where it won't cause electrical shorts or touch the exhaust manifold.

Check the throttle for free movement. There is significantly *less* accelerator pedal movement with the Weber installed. Have a helper verify that the throttle plates open and close fully as you press the accelerator.

Fuel Line—Reconnect the fuel line to the carburetor. Install a low restriction, in-line fuel filter for added insurance. Don't confuse the *fuel inlet nipple* with the *vent nipple* on the float bowl. The vent nipple is larger and has more serrations on it. The large hose from the charcoal canister connects to the vent nipple. If the car doesn't have a charcoal canister, seal the vent nipple with a short length of plugged hose or a suitable rubber cap.

Electrical Connections—There are two electrical fittings on the DGAV. One controls the idle fuel-cutoff solenoid, the other the electric choke. Using the wire harness supplied with the kit, connect both fittings to an *ignition-switched 12 volt* (v) source. The stock Lucas fuse block has a spare spade connector at its top, rearmost corner.

To verify the correct wire, use a test meter or light to determine 12v controlled by the ignition switch. When you find the right wire, connect the harness wires from the carburetor to the spare spade connector. Route the wires from the carburetor around the back of the valve cover to the fuse block.

Coolant, Gas Cap, Battery—Refill the radiator, refit the gas cap and reconnect the battery.

Air Cleaner—Installation materials are packed inside the air cleaner. Install the plastic elbow fitting to the air cleaner base. Bolt the base to the carburetor using the four bolts and spring washers. Connect a piece of hose between the elbow and the engine breather nipple.

Oil the air cleaner element and fit it in place, then assemble the rest of the air cleaner. Carefully check for adequate hood clearance before closing the hood.

Tune—Attach the new Tune-Up Specifications decal next to the original decal in the engine compartment. Reset the timing to the new specification.

Start the engine and listen carefully for air leaks. Check for fuel leaks. Stop the engine. Next, adjust the idle speed and idle mixture. Back out the idle speed screw until it doesn't touch the throttle bellcrank. Then screw it in until it just begins to touch the bellcrank. From that position, screw it in one full turn.

Start the engine and let it warm up. Stop the engine, remove the air cleaner and confirm the choke is open all the way. Reattach the air cleaner, start the engine and turn the idle mixture adjustment screw in (clockwise) until the engine begins to run rough. Slowly back it out (counterclockwise) until the engine idle is smooth. Do not back the screw out any farther than is necessary to get a smooth idle.

Because the engine now has an automatic choke, alter your starting procedure in cold weather. Depress the accelerator pedal, slowly, twice, to close the choke butterfly. Then, start the engine. Allow it to fully warm and then enjoy this conversion.

DCOEs ON AN ALFA ROMEO

The car that really put Alfa Romeo on the sports car map in the U.S. was a 1957 factory hotrod called the Giulietta Veloce. The original Veloce required 3000 rpm to get away from a stoplight, but would exceed 100 mph in utter comfort. It had a 78-CID, 90-HP in-line 4-cylinder that redlined at over 7000 rpm. The engine featured wild camshafts, forged Borgo pistons, and twin DCO Weber carburetors that took up almost as much room in the engine compartment as the engine itself.

The charisma of those early Veloces, particularly the aesthetics of the Weber carburetors, was so potent that Alfa began using the Veloce designation in much the same way our forebears used salt: to cover up the rotten taste of aging meat. Indeed, within 10 years, Alfa had applied the *Veloce* designation so generally to its cars that it had lost all meaning. In Italian, it means "fast."

The joy of that original Giulietta Veloce indelibly endeared DCOE Webers to Alfa owners. There is another reason, too. In 1969, Alfa turned to SPICA mechanical fuel injection (F/I) to meet U.S. emission laws. The F/I pump is driven off the crankshaft by a toothed belt. It was one of the more advanced fuel delivery systems available for passenger cars at the time. The U.S. Alfa distributor was very concerned about Alfa owners tampering with the emission controls and fuel metering of the fuel injection. So it was circumspect on do-it-yourself repair and tuning of the SPICA system.

Consequently, many Alfa owners removed the expensive SPICA unit and replaced it with a pair of 40 DCOE Webers. The fantastic performance increases claimed for these Alfas is probably due more to the poor state of tune of the original SPICA unit than to the virtues of Weber.

Putting Webers on an Alfa is a popular conversion, but it's not just a bolt-up operation. Alfa owners are indebted to John Shankle of Shankle Engineering for designing the kit for the job. John's kit is

Shankle kit is economical way to install dual DCOEs on four-cylinder Alfa engine with stock SPICA fuel injection.

offered through several sources as well as its home: Shankle Engineering, 9135-F Alabama Ave., Chatsworth, CA 91311. The basic Shankle conversion kit is the one installed here. The pair of 40 DCOE Webers were supplied by Redline.

Before giving details of the Weber conversion, consider the cost of the SPICA parts being removed. Don't casually toss them into a corner of the garage. The fuel-injection pump alone is worth about $700. The fragile temperature sensing tube that snakes between the manifold and the fuel-injection pump is about $200. Each fuel injector costs about $50. In all, the parts being removed during this procedure total well over $1,000 (at this printing). If all are serviceable, save and carefully pack them away. Use tape or rubber caps to seal the ends of the fuel lines and close all openings on the fuel-injection pump.

REMOVING FUEL INJECTION

Preparation—Take inventory of the parts of the Weber kit and become familiar with them. The kit utilizes much of the stock fuel-injection system and so keeps the new part count (and cost) to a minimum. The crux of the Shankle kit is the pair of thick aluminum spacers that mate the Webers to the Alfa intake manifold without modification.

Engineering of the kit is thorough. Alfa's stock fuel pump, located near the fuel tank, delivers fuel at approximately 30—40 psi. This pressure would easily overcome the needle valves in the Weber's float bowls and flood the engine with raw fuel. Shankle introduces a slight restriction in the fuel-return line to create a relatively low-pressure fuel supply.

143

Air-filter assembly lifts free as a unit.

Throttle rods are snapped free with screwdriver.

The supply is trimmed further by a diaphragm-type pressure regulator.

An alternate to this fuel supply method is to disconnect and remove the high-pressure electric fuel pump. Then, plumb and wire in a low-pressure (8—12 psi maximum) solid state electric fuel pump. These pumps are readily available; Facet is one manufacturer.

Disconnect Battery—Disconnect battery's positive lead. Several wires to the fuel-injection pump will be disconnected and taped up. Disconnecting the battery ensures protection from sparks during the conversion.

Radiator—Drain the radiator by loosening the radiator cap and then disconnecting the lower radiator hose. Reassembly is an excellent time to renew the coolant.

Air Filter—Remove the air-filter assembly. Two fabric straps hold it to the top of the manifold, and the rubber hoses between the filter and manifold are clamped between them on both ends.

F/I Linkage—Remove the two throttle rods that operate the fuel-injection pump bellcrank and the throttle bellcrank.

F/I Fuel Lines—Lubricate with penetrating oil, then loosen the four fuel lines at the injectors. Carefully pull the lines aside just enough to get at the injectors. Label each line to identify which cylinder it goes to.

Thermostatic Actuator—Use a 7mm socket to remove the *thermostatic actuator* from the manifold. Slowly pull it just free of the manifold. As noted earlier, this is an expensive piece, so handle it with care.

Injectors—Use a 17mm socket to remove the *filter holder/adapter* to each fuel injector. Remove each fuel-injector nozzle using a 19mm deep-well socket. Note: There is a flat filter disc in each nozzle; don't lose it. Place each nozzle and its hardware in a separate plastic bag for safekeeping.

F/I Pump Removal—Extracting the SPICA fuel-injection pump requires patience. The procedure here is representative, but not exhaustive. Later SPICA applications in Alfa sedans and coupes

Lubricate fuel-injection lines before loosening fitting to ensure that lines are not twisted or broken during disassembly.

Use two wrenches on fuel line fitting as shown to ensure desired joint is loosened.

Pull fuel-injection lines free just far enough to provide access to injectors.

Thermostatic actuator attaches to manifold with 7mm nuts.

Fuel-injector-filter holder/adapters are removed next.

Fuel injectors should be immediately stored in safe place.

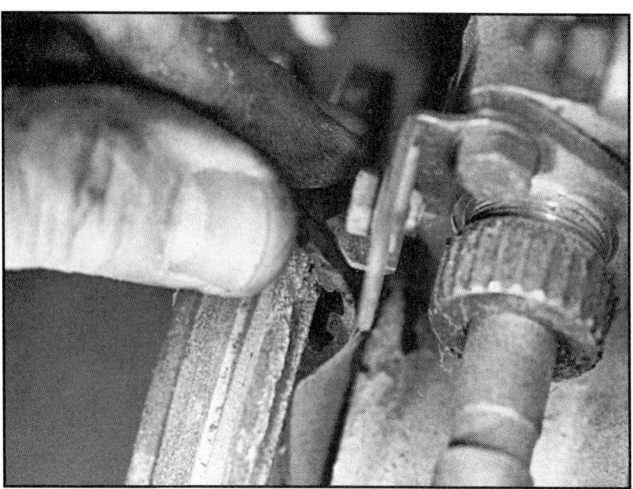
Toothed belt for driving fuel-injection pump is tight fit between crank pulley and block.

Two small bolts hold bracket that supports thermostatic actuator line.

(post-'74) used the F/I pump-drive gear to turn an air pump driven by another toothed belt. Consequently, on those engines, the air pump and its driving belt must be removed to get to the F/I pump and its driving belt.

The F/I pump's-drive gear changed in these applications. The driving belt won't just slip off the F/I pump-drive gear as in this procedure. The gear must be pulled off. All normally that has to be done is to remove the securing nut and then apply gentle persuasion with a flat-blade screwdriver to the back of the gear. It usually comes right off. If not, use a suitable puller.

F/I Pump Drive—Completely remove the lower radiator hose to gain clearance to the fuel-injection pump drive belt. Remove the distributor cap to give you more working room.

Remove the three 10mm nuts that attach the fuel-injection pump's drive-belt cover to the engine. It's likely you won't be able to see the nut closest to the crankshaft pulley. Use your fingers to locate it. Lift the cover free of the engine. Remove the fan belt by loosening the alternator.

Removing the fuel-injection pump drive belt is a challenge. Begin by carefully slipping it forward, off the fuel-injection pump gear, using your fingers

Fuel line supports are almost impossible to see; find them with your fingers and then apply wrench.

Removing thermostatic actuator at fuel-injection pump.

Fuel lines come off in order, back to front.

Remove wire from starter to cold-start solenoid.

or the flat side of a screwdriver blade. (This applies only to pre-'74 engines.) If using a screwdriver, don't *notch or cut* the belt. It doesn't stretch, so simply slip the belt forward off the gear. Work at both the top and bottom of the gear until the belt is free.

A protective boss around the crankshaft pulley has just enough clearance for the belt to thread through and be removed. Carefully twist the belt so it passes between the boss and the crankshaft pulley. Now, remove the two small bolts supporting the thermostatic actuator line below the manifold.

Remove F/I Lines—Remove the 10mm nuts holding the fuel lines to the underside of the intake manifold. Find these nuts with your fingers and then remove them. Save the notched supports that come free with the nuts.

Remove the two screws holding the thermostatic actuator to the fuel-injection pump and remove the entire actuator assembly. Be careful not to kink its copper tubing. To keep the pump dirt-free, place tissue paper in the hole where the actuator came out of the fuel-injection pump.

Begin at the back of the fuel injection pump and work forward. Use a 17mm open-end wrench to remove the fuel injection lines from the pump. Tape the

Remove bolts that attach fuel-injection pump to its support brackets.

Four studs attach fuel-injection pump to crankcase.

Blanking plate is thick. More fragile unit would be bent by oil pressure in line to fuel-injection pump.

Front castings of intake manifold are removed next.

ends of the lines, then tape the lines together in pairs (1-2 and 3-4) for storage.

Remove F/I Pump—Remove the wire from the fuel-injection pump that routes to the starter. Tape the wire and secure it out of the way. Remove the remaining wires to the fuel-injection pump and tape them out of the way.

The pump is supported at its top with two brackets. Remove all the fasteners attaching the brackets and remove the brackets. Remove both flexible fuel lines to the fuel injection pump.

You're now ready to remove the fuel-injection pump. Remove its four 13mm attaching nuts at the engine's crankcase and then lift out the pump. Three of these bolts are accessible, but the fourth—to the front of the car—is a challenge. This bolt is more accessible if the fuel-injection pump gear is removed. But you can remove the bolt with the correct tool. Buy a 1/4-in. drive, 13mm *flex socket* and use it to extract the bolt. Then lift out the F/I pump.

Install F/I Blanking Plate—A sturdy aluminum blanking plate, gasket and O-ring are included in the kit to close off the oil passage to the F/I pump. Secure the O-ring to the blanking plate with some heavy grease and place the gasket and plate on the studs for the F/I pump. Use the original nuts to attach the plate.

Remove Throttle Plates—The next two steps strip the intake manifold and make it ready for the Weber adapter plates. Remove the nuts holding the casting for the intake manifold throttle plates.

Pull free all the *idle-circuit* rubber hoses on the manifold. Remove the studs for the intake manifold throttle plate casting, and the two small studs for the thermostatic actuator. Save all the studs.

INSTALL DCOEs
Prepare Throttle Bellcrank—First, remove the bellcrank. Release the straight end of the throttle-return spring on the manifold. Bend back its lock tab, then remove the nut that holds the bellcrank to the manifold. On SPICA-injected Alfas 1969—'79, grind off the *swaged*—stamped—end of the inner ball-pivot and remove the ball-pivot from the bellcrank lever arm. The inner ball-pivot is the one that faces the engine. On later

Remove manifold attaching studs, being careful not to damage threads. Save all studs.

Throttle bellcrank nut is removed, then bellcrank.

Outermost ball is removed from bellcrank on early model Alfas.

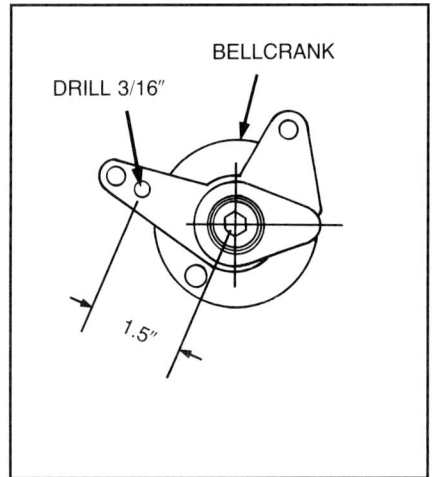
Bellcrank needs a repositioned ball for Weber carbs. Drawing courtesy Shankle Engineering.

New ball with connecting link attached.

cars, it's not necessary to remove the inner ball-pivot. Save the ball-pivot. If reinstalling the SPICA system, the ball-pivot can be brazed to the bellcrank lever arm. Similarly, remove the outer ball-pivot from the end of the bellcrank lever. Save it, too.

Drill a 3/16-in. hole in the bellcrank lever arm 1.5-in. from the center axis of the bellcrank. Attach the new ball-pivot supplied in the kit. Face the ball end away from the engine. Reinstall the bellcrank to the manifold.

Plug Injector Holes—Prepare the four blanking plugs that replace the fuel-injection nozzles. Coat their threads with a gasket sealer, then screw them snug into the manifold.

Plug Thermostatic Actuator—Install the blanking plate with O-ring for the thermostatic actuator.

Adapt Intake Manifold—First, plug the vacuum line from intake manifold to the air filter's hot-air regulating canister on models so equipped.

Inspect the gasket on the inlet manifold. Use a silicone gasket material to repair any tears in it, or renew it. Use the eight Allen-head bolts supplied in the kit to install the two adapter plates to the

Installing fuel-injection nozzle blanking plug.

Installing thermostatic-actuator blanking plate.

Adapter plates go on next.

Shankle soft mounts are plastic with large O-rings on both sides.

Dream come true; first glimpse of *Weberization*.

manifold.

Install Soft Mounts—The soft mounts for the Webers are a plastic flange with large O-rings on either side. Mount the O-rings to the flanges. Use some heavy grease to hold them in place on the adapter flanges, if necessary. Then, slip the flanges onto the adapter-plate mounting studs.

Install DCOEs—There is a spring-loaded adjustable joint used between the throttle shafts of the two Weber carburetors. The lever on the throttle shaft of the rear-most carburetor must slide *between* the adjusting screw and its spring-loaded backup on the front carburetor. This must be done while simultaneously putting the carbs on the adapter plates.

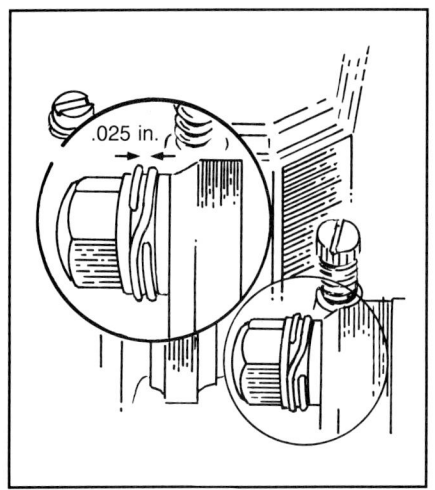

Gap between coils of lock washer is necessary for correct mounting. Drawing courtesy Shankle Engineering.

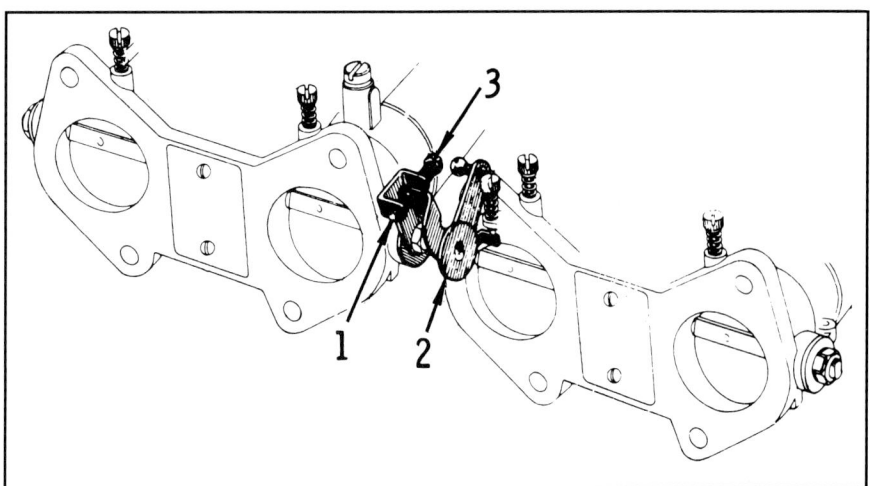

Carburetor-balance adjustment as seen from manifold side of carburetors. Drawing courtesy Shankle Engineering.

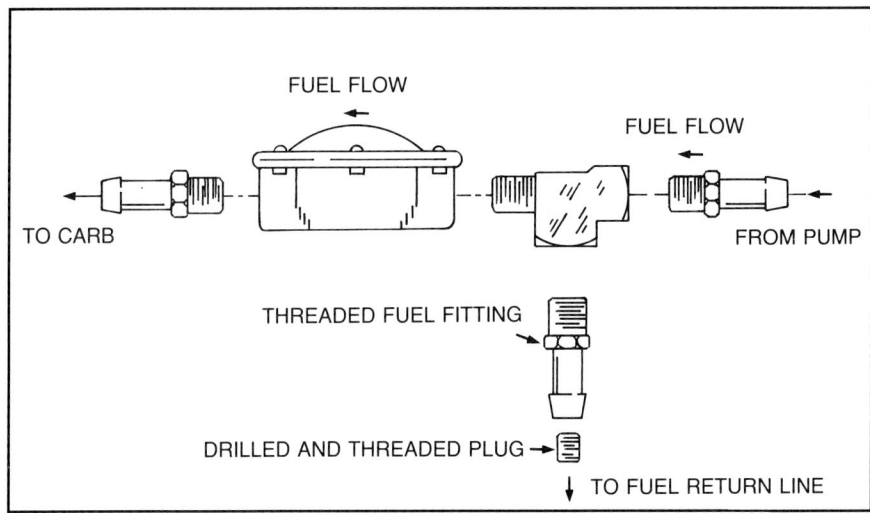

Regulator ensures correct fuel pressure at carburetors, and retains stock-configuration fuel-return line. Drawing courtesy Shankle Engineering.

This feat of dexterity may require several tries before you get the joint assembled and the carburetors on their mounting studs. An extra set of hands is helpful. Double-coil lock washers are used to provide some "jiggle" for the soft-mounted carburetors. These special lock washers are tightened so there is 0.025-in. gap between the coils.

It's impossible to see the gap on the lock washers installed to the mounting studs under the carburetors. So, tighten one of the top lock washers until it's completely compressed. Then count how much of a turn is needed to back off the nut so the lock washer has its 0.025-in. gap. Repeat the procedure, blind, for the bottom-mounting nuts.

Attach Bellcrank—Connect the link between the bellcrank on the intake manifold and the Weber bellcrank. Check that when the bellcrank is closed, the carburetor throttle valves are closed. And when it rotates fully open—press the accelerator to the floor—the valves are fully open. Adjust the length of the actuating rod between the carburetors and bellcrank by screwing in or out the lower ball-pivot socket. If complete opening and closing of the throttle valves can't be done with this adjustment, adjust the bellcrank stop screws directly behind the bellcrank.

Air Filters—These are *essential* for street-driven applications. Fit air filters, which are part of the complete Shankle kit or available from various Weber specialists. Check the clearance to the fenderwell before ordering air filters.

Manual Synchronization—If the Webers you're installing have a screw-in plug for the progression holes, unscrew the plug. The plugs are next to the idle jet screws. You can then see the position of the throttle plate by looking through the progression holes. Adjust the spring-loaded coupling so the throttle plates of both carburetors are the same relative to the same progression hole.

If there isn't access to the progression holes because a swaged plug is used in-

Fuel-regulator installation when stock fuel pump is retained.

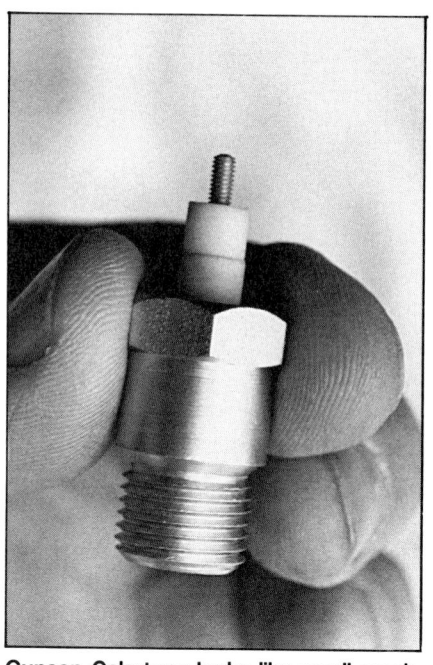
Gunson Colortune looks like small spark-plug, is used to check mixture richness.

Colortune can be read in broad daylight using stovepipe setup that's provided in kit. Mirror pivots atop stovepipe for easy, safe viewing.

stead of a screw-in plug, you can still do a fairly close manual balance of the two carburetors. Undo the idle speed adjustment screw so the throttle plates seat against the throttle bore.

Next, back the spring-loaded adjusting screw out so it no longer presses against the lever of the front carburetor's throttle shaft. Then, very carefully turn the adjusting screw in until the front carburetor's lever just begins to move. Back the screw out just enough so there is no loose play, and the screw is not pressing against the lever enough to move it. Repeat the procedure again to verify that the screw just contacts the lever, but does not move it. Then, turn the idle speed screw in about one turn.

Fuel Supply—If the stock fuel pump is used, the pressure regulator included in the conversion kit *must* be used. As noted before, the fuel is delivered from the stock pump at approximately 30—40 psi. The stock system returns excess fuel to the tank. By introducing a restriction in this loop just downstream of the carburetor tap, a low-pressure supply of fuel is available for the Webers. It is further controlled by a diaphragm-type regulator. Assemble the fittings to the regulator and then plumb the regulator into the fuel system. Set it to supply a fuel pressure of 3.5 psi.

Coolant & Battery—Connect the lower radiator hose and replenish the coolant. Connect the battery. Turn on the ignition—the fuel pump will come on—and check the fuel fittings to be sure they don't leak. Fix any leaks and check all connections.

Final Tuning—Pump the accelerator three times and start the engine. Let it warm up at a fast idle. Do the final tuning of the carburetors as detailed in Chapter Four, pages 34-40.

I used a *Colortune* kit to check mixture strength. This kit, manufactured by Gunson in Great Britain, is distributed in the U.S. by White Eagle, Rt. 1, Box 279 E. Bernstadt, KY 40729. It is essentially a sparkplug with a clear ceramic insulator that allows you to see the actual color of the burning gases in the combustion chamber!

This kit is great for tuning the mixture on fuel-injected Alfas, and was invaluable in troubleshooting the Weber conversion. The tuning procedure is to increase idle mixture richness until you see a bright yellow flame, then lean out the mixture until the flame becomes an almost invisible dark blue. At that point, the fuel is burning with maximum efficiency. Complete instructions are supplied with the Colortune.

Virtues & Vices—The SPICA system had one virtue over the conversion. It returned to idle quickly. The Webers eventually reach a rock-solid idle speed, but take their time getting there. I couldn't detect, in everyday driving, any performance gains over the SPICA system with the Webers installed.

On the other hand, the Webers are simpler to maintain and repair. The SPICA injection-pump drive belt can break, and the engine stops RIGHT NOW. The F/I pump is robust, but is expensive to repair or replace. The stock electric fuel pump at the rear of the car is similarly unrepairable by most amateurs, and it is more expensive than aftermarket electric fuel pumps that supply fuel at lower pressure.

Finally, adjusting the stock fuel-injection system requires special factory tools and, to be done correctly, an HC/CO meter. These are items the average enthusiast doesn't own.

All in all, the dual Webers have great visual appeal and are easy to repair or maintain. But don't expect performance miracles from them in this application.

QUAD-IDFs ON THE SMALL-BLOCK CHEVY

One of the most dramatic Weber installations available on any car is a set of four IDF-series Webers on an American V8. The kit being installed on a small-block Chevy is from Inglese Induction Systems of Branford, CT.

This kit requires a significant investment of money, and is not intended for the casual hobbyist. Indeed, Inglese's instructions presume that the engine has already been stripped of its intake manifold, and any emission-control devices dealt with according to the owner's desire. If this approach is a bit daunting, then step-by-step details of removing the intake manifold are available in a variety of manuals. HPBooks' *How To Rebuild Your Small-Block Chevy* is one.

Inglese's kit comes with Webers jetted for the specific engine, and is a relatively simple bolt-on installation. Your job, once the kit is installed, is getting the carburetors properly synchronized.

Multiple downdraft Webers on 289 Cobra. Photo by Ron Sessions.

Four 45 DCOE sidedraft setup on cross-ram intake manifold for small-block Chevrolet is available from Fuelish Parts Ltd., 3120 Penbroke Road 231, Penbroke Park, Florida 33009. Dyno test comparing typical four-barrel setup to that shown yielded results in chart. Bottom-end torque was greatly improved. Another 118 HP was gained from full shot of Nitrous, resulting in near 600 HP maximum! Photo and chart courtesy Fuelish Parts Ltd.

	DYNO TEST			
	4 BARREL		I.R. CROSSRAM	
RPM	TORQUE	HP	TORQUE	HP
3000	322	317	406	478
3500	331	337	414	387
4000	341	345	420	395
4500	365	351	425	404
5000	347	362	410	422
5500	320	368	399	453
6000	286	358	349	432

Fuelish Parts Ltd. offers kits for big- and small-block Fords and big-block Chevys. This one's for a big-block Ford; four 48 IDA downdrafts on a polished manifold. Photo courtesy Fuelish Parts Ltd.

When reading the following, keep in mind the orientation of the manifold: *passenger's side* and *driver's side* is used to refer to the carburetors' placement. If you turn the manifold and become too absorbed in the work at hand, you can easily confuse sides. Remember, the distributor boss is at the back of the manifold and the thermostat is at the front.

Set Float Level—The first step is to verify the float-level setting on each carburetor. The float level is a critical part of carburetor calibration, because it affects the point at which the main circuit starts and affects the operation of the emulsion tube.

Hold the carburetor float bowl cover exactly as shown in drawing (A), so the float hangs freely like a pendulum. Two dimensions must be verified. The first, at 10mm (13/32 in.), is the distance of the float from the face of the float bowl cover when the needle is just seated. Don't compress the spring-loaded ball within the needle while taking this measurement. Bend tab *6*, if necessary, to achieve the correct dimension. The second dimension is maximum float travel. It's adjusted by bending tab *3*. The maximum free travel is 32.5mm (1-9/32 in.). Verify both dimensions for all carburetors.

Fuel Pressure Regulator—Install a fuel pressure regulator inline to limit fuel system pressure to less than 3.5 psi. Inglese's experience has shown that the IDF will not tolerate over 4.0-psi fuel pressure, which is well within the range of most fuel pumps. If system pressure exceeds 3.5 psi, the carburetors will flood and, aside from causing the engine to run poorly, create a fire hazard.

Assemble Bellcrank—Next, assemble the bellcrank as shown in drawing (B). The bellcrank uses spherical rod ends at the end of the rods. These rod ends must always be assembled with a nylon washer on either side, as shown. The large pivot bolt runs in two needle bearings, which should be carefully pressed into place using a bench vise. Protect the surfaces of the bearing and bellcrank by using wood between the assembly and the vise jaws.

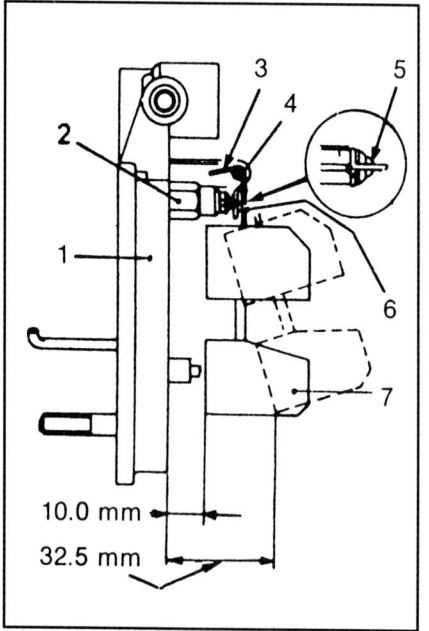

(A) Float-level adjustment must be checked on each carburetor before it is installed. Two dimensions must be measured. Drawing courtesy Inglese Induction Systems.

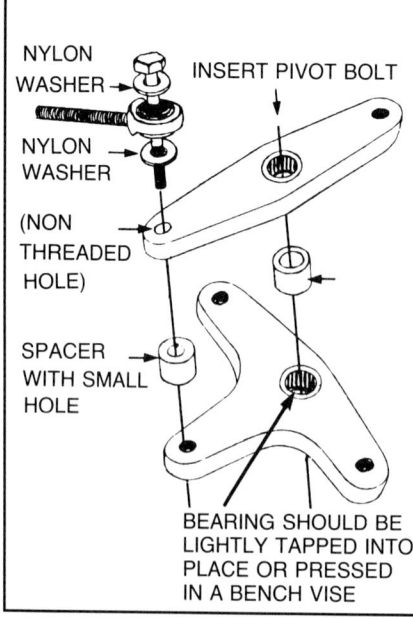

(B) Assemble bellcrank. Drawing courtesy Inglese Induction Systems.

(1) Attach bellcrank to manifold. Photo by Bruno Ratensperger.

(2) Attaching spring-loaded levers to passenger's side front and driver's side rear carburetors. Photo by Bruno Ratensperger.

Assemble & Install Linkage—Refer to photo (1): Note the short linkage rods installed in place on the mounted bellcrank assembly. The two short rods must be the correct length to seat the throttle shaft bellcranks against their stops with the idle-speed adjustment screws removed. Both short rods must be exactly the same length. To adjust them, one end of the rod has a groove in it to indicate a *left-hand* thread. Use this end to adjust the rod length so it presses the throttle shaft bellcrank against its stop, and then use the left-hand-thread locknut to secure the rod's position.

Install the two carburetors as shown in photo (1). The *male* linkages that are controlled by the bellcrank are attached the same way on both the passenger-side rear carburetor and driver's side front carburetor. Attach the long male lever as shown in photo (1). This male lever will eventually connect with the matching *female* throttle lever of the other carburetor in the longitudinal pair.

Tighten the throttle-shaft nuts lightly. If you overtighten them, they will cause the throttle bellcrank to bind against the carburetor body. Bend the locking tabs over each throttle-shaft nut to keep the nuts from vibrating loose.

Install the idle-speed adjusting screw where the arrow indicates in photo (1), but not so far that it moves the bellcrank off its stop. Note that only one idle-speed adjusting screw is used for the entire system, and that it is on the passenger's side rear carburetor.

Attach the female—*spring-loaded*—levers to the ends of the passenger's side front and driver's side rear carburetors exactly as shown in photo (2). Check the direction of the levers against the location of the carburetor fuel inlets to verify that the levers are correctly installed. Tighten the throttle-shaft nuts lightly, then secure them with the lock tabs.

To make the accelerator pedal acceptably light, remove the throttle return springs from the driver's side front, and passenger's side rear carburetors. They should look just like photos (3) and (4), respectively.

Study photo (5). Install the two carburetors with female levers (arrows) so the male lever of the carburetor pair is held between the adjusting screw and the spring-loaded stop of the female lever. Secure the carburetors to the manifold. Don't overtighten the mounting nuts.

The throttle plates of the carburetor with the female levers are probably forced slightly open and not synchronized with the the plates of the other—male—carburetor in the pair. Use your thumb to close the throttle plates of the female lever against the pressure of its spring stop, then screw in the adjusting screw on the female lever so it just touches the male lever. Verify that all the throttle plates are closed. If one is left open, the engine will idle at a very high rpm when started.

Assemble the throttle cable as shown in photos (4) and (5), and complete assembling the linkages to the carburetors. The unused bellcrank arm, photo (5), can be used for a transmission kick-down cable, if necessary.

Work the bellcrank assembly to ensure there is no binding, and that the throttle plates freely close in synchronization. Also, check that each throttle plate opens fully by pulling on the throttle cable.

Install Carbs & Manifold—To install the manifold, first fit the gaskets to the head. Inglese recommends not to use the neoprene front and rear manifold-to-block gaskets from the gasket kit. Instead, run a heavy bead of silicone gasket material along the front and rear mating surfaces where the neoprene gasket would normally seal. Let the silicone set

(3) Return spring removed from driver's side front carburetor. Photo by Bruno Ratensperger.

(4) Return spring removed from passenger's side rear carburetor. Photo by Bruno Ratensperger.

up about 30 minutes to one hour. Then, carefully lower the manifold *straight down* onto the engine. Secure it in place with bolts and, after the silicone is fully set—note the instructions on the tube—trim off the excess with a razor blade.

Install Distributor—This carburetion setup permits only a small-diameter distributor. A stock HEI distributor will not fit, and vacuum-advance distributors are not recommended by Inglese. He recommends a high-quality aftermarket mechanical-advance distributor.

Even with a small-diameter distributor, it is possible that the distributor-cap screw or external condenser can be situated so it will foul against the carburetors. Rotate the distributor so that 36—38° of total advance can be achieved. That is, the distributor should be free to rotate at least 45° without interference. Then rewire the distributor cap to match drawing (C). Set the static—initial—ignition timing.

First Start!—Disconnect from the distributor *and* ground the primary coil wire so there is no spark to the plugs. Verify that the fuel-pressure regulator is set for 3.5 psi. Have a friend standby to watch for fuel leaks. Feed fuel to the carburetors. If the installation uses an electric fuel pump, simply turn on the ignition. If a mechanical pump is used, crank the engine to fill the float bowls with fuel.

If there are no fuel leaks, reconnect the coil wire. Press the accelerator pedal to the floor twice and release it, then start the engine. Because there is no idle adjustment, use the accelerator pedal to keep the engine running. The engine will probably cough slightly on start-up because it is cold, but will clear in about a minute. If the engine doesn't start immediately, or mis-fires, recheck static ignition timing.

Set Ignition Timing—Static timing is of secondary importance to the performance level of a small-block with four IDFs, but setting maximum advance is critical. It *must* be set correctly before the carburetors are tuned. The total advance should be 36—38° at maximum rpm.

Check for maximum advance by putting *timing tape* on the crankshaft vibration damper/pulley and using an accurate stroboscopic timing light. If any wires foul against a carburetor as you rotate the distributor while setting timing, shut off the engine and rewire the distributor cap. Move the sparkplug wires *back,* or *ahead* one terminal hole so you can achieve the correct timing. If you can't get enough advance for example, move each sparkplug wire ahead—counterclockwise—one terminal. For more retard, move each wire back—clockwise—one terminal.

Synchronize Carbs—With timing set and the engine warm, the next step is to balance carburetor airflow. In this installation, check airflow at each velocity stack. There are a number of acceptable airflow indicators that can do this job. The Unisyn is one. In any case, read the instructions of the measuring unit you're using. You may have to adjust it so it will correctly read flow rates of your engine.

As shown in photo (6), check flow of the passenger's side rear carburetor—it's the one with the idle speed adjustment. Measure only one throat of each carburetor. The front throat is suggested; the rear throat is adjusted later.

Center the indicator over the velocity stack so its own venturi is concentric with the carburetor's venturi. As shown, set the idle-speed adjustment screw so the engine idles as slow as possible without stalling. Then adjust the Unisyn, or whatever device you're using, so its indicator falls in the center of its calibrated tube. This *centered* flow indication is the baseline measurement that the other three carburetors will be adjusted to match.

Before proceeding to the other carburetors, remove the Unisyn, rev the engine briefly, let it return to idle, and

(5) Carburetor pairs install with male and female levers engaged. Photo by Bruno Ratensperger.

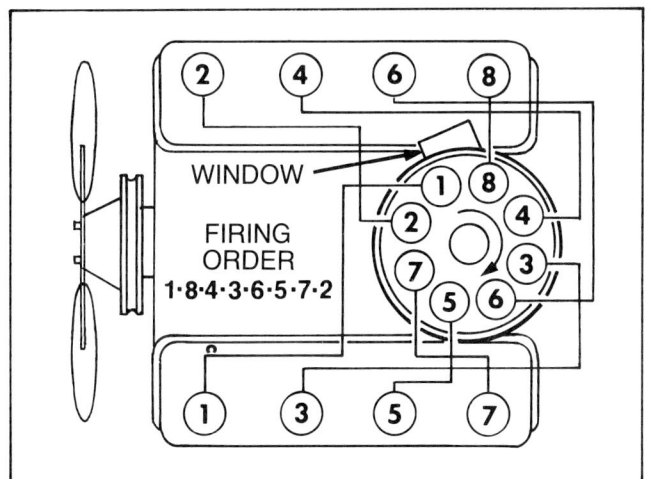

(C) Distributor wiring on small-block Chevy. Drawing courtesy Inglese Induction Systems.

(6) Initial flow measurement using Unisyn. Photo by Bruno Ratensperger.

recheck flow. The procedure of revving the engine to clear it and then rechecking flow must be done after every adjustment on every carburetor.

With the passenger's side rear carburetor adjusted, move to the driver's side front carburetor, photo (7). As shown, use a 3/8-in. wrench to turn the hex-link until the Unisyn reads baseline flow. Rev the engine and recheck, then tighten the jam nut on the hex-link.

Adjust the driver's side rear carburetor next, photo (8), using the speed adjusting screw of the female—spring-loaded—lever. To increase the rate of flow, back out the screw. To decrease flow, turn in the screw. Obtain the centered reading on the Unisyn, then recheck.

The final carburetor on the passenger's side front is adjusted using the female adjusting screw as on the third carburetor, photo (9).

After all four carburetors have been adjusted, you may have to reset the system idle speed by turning the idle-speed adjusting screw on the passenger's side rear carburetor. Recalibrate the Unisyn to center its indicator on the passenger's side rear carburetor. Then, once again, check flows of the remaining three carburetors. They should agree exactly. Rev and recheck.

Adjust Mixture—Idle mixture richness

(7) Adjust second carburetor. Photo by Bruno Ratensperger.

(8) Adjust third carburetor. Photo by Bruno Ratensperger.

is adjusted using screw (A) for each throat, photo (10). For initial adjustment, stop the engine and turn in each screw on every throat so it just seats lightly. Then back it out 1/2 turn. Verify that all idle mixture screws are 1/2-turn out from seated, regardless of the orientation of the screw slot.

Start the engine and back out each screw 1/4 turn, then rev the engine to clear it. Note idle speed. Keep track of the total numbers of turns you have moved the adjusting screws for reference. Keep backing out the screws equally, 1/4 turn at a time, clearing the engine and then noting the idle speed.

The engine will continue to increase its idle until it reaches a point at which idle speed does not rise, but begins to drop. Turning the screws out farther will cause the idle to drop even more and the engine will begin to run rough. The desired setting is that point where idle speed won't increase, but the engine doesn't begin to run rough. It may be necessary to repeat this procedure again to verify that the correct idle mixture has been set.

Balance Barrel Airflow—After you're sure the mixture is right, recheck the front venturi of the passenger's side rear carburetor for its baseline reading. Adjust the Unisyn as necessary to establish a baseline reading. Then, measure the airflow in the other venturi of the *same* carburetor. If the readings are within one full unit on the Unisyn, no adjustment is necessary. If outside this tolerance, increase the airflow of the lower reading carburetor barrel by adjusting the air-bleed screw, (B) photo (10).

To adjust the air-bleed screw, loosen its locknut and turn out the screw to increase airflow; or turn in to reduce airflow. Retighten the locknut when flow in the lower-reading barrel is the same as the other barrel.

Check and adjust the balance between the barrels of the other three carburetors the same way.

Repeat As Necessary—After making all adjustments following the above procedure, re-adjust idle speed so it is exactly what you want and repeat the *entire* procedure, beginning at square one. Adjust airflow through one venturi of each of the carburetors, then adjust idle mixture. Then, if required, re-balance the remaining venturi of each carburetor.

Perfectionists will insist on performing the procedure a third time, before going for a drive. And remember, once all adjustments are established, leave the carburetors alone!

(9) Adjust fourth carburetor.

(10) Set idle-mixture screw (A) and balance airflow between venturis with air-bleed screw (B).

Inglese Troubleshooting Guide

- **Black Smoke:**
  Mixture screws turned out too far
  Float level set too high
  Excessive fuel pressure to carbs
  Jets too large
- **Carbs "sneezing" under light to moderate throttle:**
  Timing not advanced enough
  Mixture screw(s) not open enough (lean condition)
  Too small an idle jet
- **Dull popping from exhaust:**
  Mixture screw(s) too lean. Open suspected mixture scew(s) 1/8-turn

- **Backfiring from exhaust under acceleration:**
  Fouled sparkplug, usually caused by over-rich mixture
- **"Flat spot" at driveaway, or up to approximately 2000 rpm:**
  Idle jets too small
- **Engine goes "flat" before redline:**
  Air-correction jets too large
- **Poor power thoughout driving range:**
  Timing not advanced enough
  Main jets too small

- **Exhaust "popping" on deceleration:**
  Exhaust leak. Usually more pronounced with independent runner (IR) induction and four-tube header(s)
- **Fuel smell during hard acceleration:**
  Excessive reversion—standoff. Common with IR induction and camshaft with excessive overlap. Control by limiting overlap and installing air cleaners

(Guide courtesy Inglese Induction Systems.)

PRODUCTION JETTING TABLE

Understandably enough, Webers that appear on production cars are far too numerous to list here. The intents of this table are to offer standard jettings for selected carburetors and to relate them to an *engine displacement*.

By seeing what jettings the manufacturers use on various models, especially the DCOE and IDF series with their range of variables, you should have a starting point. Refer to the tuning chapter on this book for detailed methods of calibrating your application.

Comprehensive production jetting tables are available from Weber dealers. They are contained in a book (Weber P/N 95.0000.15), which lists jet sizes for all factory (OEM) installations, but not detailed tuning procedures.

This table is arranged by CARBURETOR TYPE and ENGINE DISPLACEMENT.

CARBURETOR	ENGINE SIZE (cc)	PRI/SEC VENTURI	AUX. VENTURI	EMULSION TUBE	MAIN JET	AIR CORR.	IDLE JET	PUMP JET	NEEDLE VALVE	VEHICLE
32 DATRA	1290	22	4.00	F30	1.10	1.90	0.50	0.50	1.50	FIAT X1/9
36 DCD 7	1089	23	4.50	F23	1.15	2.00	0.45	0.70	1.75	FIAT 1200 H
3/40 DCNF 19	2418	32	4.50	F24	1.25	2.20	0.55	0.50	1.75	FERRARI DINO 246
4/40 DCNF 45	2926	32	4.50	F26	1.35	2.20	0.55	0.45	1.75	FERRARI 308/GT4
1/45 DCOE 13	1266	38	5.00	F16	1.60	*	0.45-F9	0.50	1.75	MINI COOPER S
2/48 DCOE 18	1559	30	4.50	F11	1.10	1.55	0.45-F9	0.40	1.75	ELAN S4-SE
2/45 DCOE 14	1570	30	4.50	F16	1.35	2.20	0.50-F8	0.35	1.50	ALFA GTA
2/40 DCOE 32	1779	32	4.50	F9	1.30	2.00	0.50-F8	0.35	1.50	ALFA 1750
2/40 DCOE 18	1800	30	4.50	F9	1.35	2.20	0.45-F9	0.40	1.75	VW RBT/SRCO
2/40 DCOE 88	1999	34	4.50	F16	1.25	1.70	0.55-F8	0.40	2.00	BMW 2000
2/40 DCOE 2	2000	33	4.50	F16	1.20	1.90	0.50-F8	0.40	1.75	BMW 2002
2/40 DCOE 18	2200	33	4.50	F11	1.25	1.75	0.45-F9	0.40	1.75	CHRYSLER K2.2
3/40 DCOE 18	2400	28	4.50	F11	1.20	1.70	0.50-F9	0.40	1.75	DATSUN 240/260Z
2/45 DCOE 9	2500	36	4.50	F11	1.30	1.65	0.55-F8	0.45	1.75	FIERO
3/40 DCOE 18	3400	30	4.50	F11	1.20	1.80	0.45-F9	0.40	1.75	BMW 3400
3/45 DCOE 2	4200	38	3.50	F2	1.65	1.90	0.65-F8	0.40	1.75	JAGUAR XK-E
32/36 DFAV	1997	26	4.50	F6	1.35	1.70	0.45	0.65	2.00	CAPRI 2000 GT-CA
		/27		F6	1.40	1.80	0.50			
2/40 DFO 2/3	1897	32	4.50	F2	1.50	1.90	0.50	0.40	1.50-	OPEL REKORD
34 DGAS 8A	2551	24	4.50	F50	1.22	1.80	0.45	0.55	2.50	EURO 2500-CM
38 DGAS 6C	2994	27	4.00	F50	1.42	1.85	0.45	0.55	2.50	CAPRI 3000
32/36 DGAV	1599	26	3.50	F50	1.35	1.70	0.55	0.50	2.00	CORTINA
		/27		F6	1.45	0.70	0.45			
32 DGV 7A	1298	23	3.50	F50	1.25	1.80	0.55	0.55	2.00	ESCORT GT
		/24		F6	1.30	1.95	0.50			
34 DMSA1/100	1755	25	3.50	F7	1.40	1.65	0.50	0.50	1.75	FIAT 124
		/27		F7	1.45	1.55	0.60	0.50	1.75	
2/40 IDA 3C/1	1991	30	4.50	F26	1.25	1.80	0.55	0.50	1.75	PORSCHE 911L
2/46 IDA 3C/1	1991	42	4.50	F24	1.70	1.45	0.70	0.50	1.75	CARRERA 6
2/36 IDF	1200	27	4.50	F11	1.10	1.25	0.50	0.40	1.75	AIR COOLED VW
2/40 IDF 42	1266	28	4.50	F11	1.10	1.30	0.50	0.40	1.75	ALFA ROMEO
2/40 IDF	1600	28	4.50	F11	1.15	1.15	0.50	0.50	1.75	AIR COOLED VW
2/44 IDF 50	1800	36	4.50	F11	1.35	1.75	0.50	0.50	1.75	VW DUAL PORT
2/44 IDF	2000	36	4.50	F11	1.15	1.70	0.50	0.55	1.75	AIR COOLED VW
2/40 IDS 3C/1	1991	32	4.50	F3	1.30	1.80	0.55	0.50	1.75	PORSCHE 911S
2/40 IDT 3C/1	1991	27	4.50	F2	1.10	1.85	0.50	0.50	1.75	PORSCHE 911T
4/40 IF 3C	4942	32	5.00	F82	1.50	2.10	0.50	0.45	1.75	FERRARI BB 512

*CLOSED: Air corrector was sealed.

STANDARD TUNINGS FOR POPULAR CARBURETORS

This table is for inital setup of the more popular Weber carburetors, with the most frequently used venturi sizes. The calibrations shown will allow you to run the engine and drive the car in order to verify jetting. These are BASELINE FIGURES ONLY. They are based on production jettings (not theoretical figures) for these carburetors on actual vehicles, and always with CORRECTLY SIZED VENTURIS for the application.

Most important are the emulsion tube selections, which should be used unless:

1. The calibration for your application has been determined by a tuning professional.
2. Your vehicle came with Webers and the tubes installed are of the original F designation for that car.
3. You are installing a conversion kit calibrated for your car by the kit manufacturer

Refer to the altitude compensation table, page 166, for operation above 5000ft, and main jet changes necessary for maximum performance.

CARBURETOR	MAIN JET	AIR CORR.	E-TUBE	IDLE JET	IDLE AIR	ENGINE SIZE
DCNF 32 or 36	1.20	2.00	F36	0.45	1.50	1.3L-2.0L
DCNF 40	1.30	2.10	F24	0.50	1.40	2.0L-3.0L
DCNF 40 or 44	1.30	1.90	F36	0.50	1.55	(1)
DCNF 42 or 44	1.35	1.60	F25	0.60	1.50	(1)
DCOE	1.35	1.55	F9/F11	0.50	F8/F9	(2)
DFAV Progr.	1.40	1.65	F6	0.45	1.50	(2),(3)
(Secondary)	1.45	1.75	F6	0.50	0.70	
DGAV Progr.	1.35	1.75	F50	0.50	1.50	(2)
(Secondary)	1.30	1.45	F66	0.50	0.70	
IDA	1.55	1.80	F14/F24	0.60	0.80	(2)
IDF	1.20	1.75	F11	0.50	1.15	(2)
IDT	1.10	1.80	F1/F2	0.50	1.10	(2)

Notes:

1. With multiple carburetors; one venturi per cylinder.
2. All engines, assuming correctly sized venturi for application.
3. For simultaneous throttle opening versions of this carburetor, use primary settings only.

DCOE AND IDA IDLE JET SELECTION

With the exception of the DCOE and IDA series carburetors, the idle circuitry on all Webers incorporates a replaceable idle jet with a calibrated fuel orifice and uncalibrated air entry orifices. The air for idle emulsion is drawn through a calibrated bushing usually pressed or drilled into the carburetor casting.

The DCOE series carburetors utilize replaceable idle jets that have calibrated fuel *and* air orifices. They are stamped with a fuel orifice size, followed by an "F" number that identifies the air orifice area; the jets are drilled with one or two air openings of various sizes in addition to the fuel orifice. This table lists the range of fuel orifice sizes available for each "F" number.

Remember that Weber doesn't assign "F" numbers sequentially rich or lean. According to experienced Weber tuners, DCOE idle jet "F" numbers can be chosen from this sequence:

(RICH) F6-F12-F9-F8-F11-F13-F2-F4-F5-F7-F1-F3 (LEAN)

The fuel orifice size most commonly used as a starting point is 0.50mm, which is available for all "F" numbers except F4 (use 0.45mm), F19 (0.45mm only), or F24 (0.45mm only)

DCOE IDLE JET TABLE

	F#	AVAILABLE FUEL ORIFICES	WEBER P/N
RICH	F6	0.40—0.70mm	74819.XXX
	F12	0.40—0.55mm	74825
NORMAL	F9	0.40—0.65mm	74822
NORMAL	F8	0.40—0.65mm	74821
	F11	0.40—0.70mm	74824
	F13	0.40—0.55mm	74826
	F2	0.40—0.55mm	74815
	F4	0.40—0.65mm	74817
	F5	0.40—0.70mm	74818
	F7	0.40—0.55mm	74820
	F1	0.40—0.55mm	74814
LEAN	F3	0.40—0.70mm	74816

OTHER TUBES AVAILABLE

F#	Fuel Orifices	Weber P/N
F14	0.50 only	74827
F15	0.50 only	74828
F17	0.50—0.55mm	74830
F18	0.50 only	74831
F19	0.45 only	74832
F21	0.55 only	74833
F22	0.50 only	74834
F24	0.45 only	74836
F26	0.50-0.55mm	74838

IDA carburetors utilize idle jet holders, each with a calibrated aor orifice of 0.60mm, 1.00mm, or 1.20mm. The idle jets, are stamped "F10" to identify them as IDA jets, are available with fuel orifices ranging from 0.40mm to 0.80mm in 0.05mm increments.

EMULSION TUBE SELECTION TABLE

With the staggering selection of emulsion tube characteristics, and the chronological assignment of "F" numbers to identify them, this table should help with the tuning of the more popular Weber models.

Richer or leaner fuel curves are achieved by altering the diameters of the tube itself (to establish th volume of fuel in the well), its internal bore, and the arrangement of radial holes drilled into it. The initiation of main metering is determined by the position and number of holes in the top portion of the tube. This is the primary reason the float level is such a critical item on Webers; if the float level is low, the engine will run lean until the airflow is sufficient to draw fuel from the well around the tube. The reverse is true for too high a float level.

TUNING PREFERENCE	61450 TUBES DCOE, IDA & F	61455 TUBES DCD, DCZ	61440 TUBES DFAV, DGAV, IDA (3V only)
Leaner top end	F19	F8, F9, F31	F8, F16, F20
Leaner low end and throttle responce	F2, F3, F14, F15	F26, F33	F33, F34
Common usage	F9, F11, F16	F23, F26	F2, F3, F11
Richer low end and throttle responce	F7	F30	F5, F7, F21
Richer throttle responce, no change to top end mixture	F8	F13	F25
Alcohol usage	F2, F3, F4, F7, F17	F8, F10, F29 F25, F26	F2, F20, F24

WEBER SPECIAL SERVICE TOOLS

DESCRIPTION	WEBER P/N
Weber Factory Overhaul Manuals:	
DCOE	95.0022.35
DGV/DGAV	95.0044.35
DIR	95.0010.35
DCD	95.0004.35
40 & 46 IDA 3C (Three-bbl)	95.0020.35
Comprehensive Jetting tables (18 pages)	95.0000.15
Emulsion tube bore reamer (DF & DG series)	96.325.765
Auxiliary venturi puller	96.101.535
Needle & Seat Gauge (for 48 IDA, 40,46 IDA 3C)	98.014.200
Float Spring (use with 98.014.200)	98.013.800
Needle & Seat Gauge (for 48 IDA)	98.015.500
Float Gauge (for DCOE carburetors)	98.006.663
Universal (all Webers) Float Gauge	98.007.350
Weber Tee wrench (8, 10 & 12mm sockets)	98.023.200
Small Tube-style wrench (8, & 10mm sockets)	98.012.850
Spring installation tool (hook tip)	98.011.600
Pin Punch (for float pivot pin removal)	98.011.400
Lead gallery plug puller (for 3.5mm plugs)	98.008.650
Lead gallery plug puller (for 5 or 6.2mm plugs)	98.011.550
Weber screw staking tool	98.010.900
Weber Synchrometer (to synchronize multiple carbs, or analyze airflow on progressive secondary carburetors)	
For small displacement engines	STE-SK
For small and large displacement engines	STE-SK
(**Note:** will sync Weber, Hitachi, Holley, Solex, SU or Zenith carburetors. May require adaptor; contact Weber dealer)	

WEBER INSTALLATION AND MOUNTING PARTS

DESCRIPTION	WEBER P/N
DCOE Softmount Kits: (Includes all hardware)	
For 40 & 42mm DCOE	99005.140
For 45mm DCOE	99005.145
For 48mm DCOE	99005.148
For 50mm DCOE	99005.150
DCOE Softmount parts:	
Neoprene O-ring mounting plate	99005.110
Set of 4 O-rings	99005.540
Set of 4 Coil spring washers	99005.530
Nyloc lock nut	99005.507
Vibration buffer	99005.570
Captive washer	99005.571
DCOE O-ring mounting kit	99005.111
Mounting gaskets:	
40 DCOE (2 req'd)	99005.031
DGV, DGAV, DFAV	99005.068
DGV, DGAV (insulating)	99005.119
DGEV (insulating)	99005.120
Universal linkage kits:	
DGV, DGAV (cable type)	99007.116
DFAV, DFEV (cable type)	99007.932
Single DCOE	99006.104
Dual DCOE	99006.105
VW Type 1 Dual Port w/2 ea IDA or DCNF linkage	99004.900
VW Type 1 Dual Port w/2 ea IDF linkage	99004.912
VW Type 3 (all) w/2 ea ICT linkage	99007.311
Porsche 914 & VW Type IV w/IDF linkage	99004.295
(**Note:** most VW and Porsche configurations are available from Weber dealers)	
DCOE Throttle wheel	99301.366
Weber Inlet nipple	99005.233

WEBER GASKET KITS AND REPAIR KITS

CARBURETOR	GASKET KIT	REPAIR KIT	OVERHAUL KIT	PUMP DIAPHRAGM
40 DCNF	92.0077.05	92.1094.05	N/A	47407.050(c)
40 DCOE 2	92.0015.05	92.1047.05	N/A	N/A
40 DCOE 18	92.0015.05	92.1046.05	N/A	N/A
40 DCOE 32	92.0015.05	92.1015.05	N/A	N/A
40 DCOE 9	92.0015.05	92.1047.05	N/A	None
40 DCOE 13, 15, 16	92.0015.05	92.1292.05	N/A	None
48 DCOE	92.0015.05	92.1321.05	N/A	None
32/36 DFAV, DFEV	92.0105.05	92.1130.05	N/A	47407.016
40 DFAV	92.0047.05	92.1122.05	N/A	47407.027
32 DFTA	92.0171.05	92.1246.05	N/A	47407.050
32/36 DGAV, DGEV	92.0108.05	92.1137.05	92.2170.05	47407.048
32 DGEV	92.0108.05	92.1137.05	N/A	47407.048
32 DGV	92.0107.05	32.1136.05	N/A	47407.048
32/36 DGV	92.0107.05	92.1136.05	92.2166.05	47407.048
34 ICT	92.0164.05	92.1213.05	N/A	47407.016
40 IDA 3C, 3C 1	92.0058.05	92.1076.05(a)	92.2044.05	47497.010
46 IDA 3C 1	N/A	92.1063.05(a)	N/A	47407.010
48 IDA	92.0034.05	92.1050.05	N/A	None
40 IDF 13, 15	92.0092.05	92.1110.05	N/A	47407.027
40 IDF 48, 49	N/A	92.4190.05(a)	N/A	47407.027(c)
44 IDF 50, 51	92.0124.05	92.0281.05(a)	N/A	47407.027(c)
48 IDF 4, 5	92.0184.05	(b)	N/A	47407.027

Notes:
    (a) No float.
    (b) No repair or complete overhaul kit available; order individual parts as needed.
    (c) Included in the repair kit.

ALTITUDE COMPENSATION TABLE

The figures in this table represent several **main** jet sizes and the appropriate altitude correction if the "normal" jet is correct anywhere between sea level and 5000 ft. The correction is to compensate for the tendency of the metering system to go richer at high altitude as the amount of oxygen per volume diminishes.

This table is presented regardless of Weber carburetor type. For other "normal" jettings, simply select a leaner jet that is the closest to the valve differences shown here.

ALTITUDE (ft.)

0-5000	5000-6500	6600-9800	9800-13000
2.00	1.95/1.90	1.85	1.80
1.75	1.70	1.65	1.60
1.50	1.45	1.40	1.35
1.25	1.20	1.15	1.13
1.00	0.97/0.95	0.95/0.93	0.93/0.90

STANDARD JETTING

These calibrations are direct from Weber.

Carburetor Model	Part Number	Main Venturi	Aux. Venturi	Main Jet	Em. Tube	Air Corr. Jet	Idle Jet	Pump Jet	Pump Ex.	Float Valve	Choke Op.	Original Application
32 DFTA	18870.277	22/22	4.0/4.0	100/105	F22/F22	250/250	60/60	45	—	1.5	Elec.	Ford Fiesta
32 DGV 7A	18870.056	23/24	3.5/3.5	125/140	F50/F6	180/180	55/50	55	—	2.0	Man.	European Ford
32 DGEV	18870.271	23/24	3.5/4.5	120/115	F50/F50	160/120	45/45	45	—	2.0	Elec.	European Ford
32/36 DFAV 32/36 DFEV	22680.056 22680.070	26/27	3.5/3.5	137/140	F66/F50	165/160	60/50	55	—	2.0	Water Elec.	Ford Capri
32/36 DGV 5A	22680.005	26/27	3.5/3.5	140/135	F50/F6	165/160	55/50	50	—	2.0	Man.	European Ford
32/36 DGAV 33B 32/36 DGEV 33B	22680.051B 22680.033B	26/27	3.5/3.5	140/140	F50/F50	170/160	60/50	50	—	2.0	Water Elec.	Replacement Conversion
34 ADF 15/150	18890.188	24/26	4.5/4.5	120/140	F11/F75	140/180	47/90	40	—	1.75	Water	European FIAT
34 DATRA 20	18890.183	23/27	4.0/4.0	117/130	F21/F25	225/190	50/50	45	—	1.75	Water & Elec.	European Lancia
34 DMSA 1/100	18890.053	25/27	3.5/3.5	145/145	F61/F61	180/165	50/60	55	—	1.75	Man.	FIAT 124 1800
34 DMTR 21	18890.062	25/26	4.0/4.0	120/150	F30/F30	160/240	50/100	45	—	1.75	Man.	European Lancia
34 ICH	15290.027	29	3.5	165	F6	190	50	55	40	1.75	Man.	Vauxhall Van
34 ICT	15290.035	29	4.5	130	F78	160	52	50	40	1.75	None	Spec. VW Conversion
40 DCNF 12	18950.060	32	4.5	125	F24	220	50	45	—	1.75	Man.	FIAT Dino
40 IDF 13, 15	18950.057/059	32	4.5	125	F11	210	55	40	80	1.75	Man.	European FIAT
40 IDF 48, 49	18950.126/127	28	4.5	115	F11	200	50	50	55	1.75	None	Spec. VW Conversion
40 DCOE 2	19550.005	33	4.5	115	F16	150	50F9	35	55	2.0	Man.	Alfa Romeo
40 DCOE 18	19550.013/.014	30	4.5	115	F11	200	45F9	40	50	1.75	Man.	Lotus
40 DCOE 32	19550.043/.044	32	4.5	130	F9	200	50F8	35	60	1.5	Man.	European Alfa Romeo
40 DFAV 1	18950.051	28	4.5	180	F2	185	60	50	—	2.5	Water	European Ford
40 IDA3C, 3C1	31300.001/.002	30	4.5	125	F26	180	55	50	Closed	1.75	None	Porsche 911
44 IDF 50,51	18990.024/.025	36	4.5	135	F11	175	50	50	55	1.75	None	Spec. VW Conversion
45 DCOE 9	19600.017	36	4.5	145	F16	155	55F8	45	40	2.0	Man.	Alfa Romeo
45 DCOE 13	19600.022	34	3.5	130	F2	175	50F9	50	50	2.25	Man.	Special Conversions
45 DCOE 15, 16	19600.004/.005	38	5.0	125	F9	170	45F8	40	70	2.25	Man.	BMW Alpina
46 IDA3C, 3C1	31360.001/.002	42	4.5	170	F24	145	70	50	40	1.75	None	Porsche 906
48 IDA 4R	19030.015	37	4.5	135	F7	120	70F10	40	Closed	2.0	None	Special Conversions
48 IDF 4, 5	19030.016/.017	40	4.5	150	F2	180	55	50	Closed	2.0	Manual	Special Conversions
50 DCO/SP	19650.001/.002	46	5.0	180	F7	160	60F8	45	100	2.5	None	Racing Conversions
55 DCO/SP	19700.001/.002	48	5.0	180	F9	180	60F8	45	100	2.5	None	Racing Conversions

CONVERSION KITS

The following jet recommendations are the result of both dynamometer and road testing. There are, however, several ways of calibrating a carburetor to achieve the same result. The manner in which carburetors are set up on an engine such as throttle opening measurement at idle, float level and degrees of inclination, will all have a bearing on the calibration requirements as will engine modification, general mechanical condition and individual driving styles. So, use these calibrations as a starting point from which to carry out further fine tuning.

Conversion Kit Number	Application	Carburetor Quantity & Model	Main Venturi	Aux. Venturi	Main Jet	Em. Tube	Air Corr. Jet	Idle Jet	Acc. Pump Jet	Pump Ex. Valve
K009	Sprite, Midget 950, 1100	1 of 40 DCOE 2	32	4.5	140	F2	175	45F9	50	Closed
K011	Sprite, Midget 1275	1 of 45 DCOE 13	34	3.5	130	F2	175	50F9	50	50
K041	MG-A	1 of 45 DCOE 13	34	3.5	145	F16	160	50F9	50	Closed
K043	MG-B '63-'74	1 of 45 DCOE 9	36	4.5	150	F16	170	50F8	60	Closed
K044	MG-B '75 on	1 of 45 DCOE 13	34	3.5	145	F2	165	50F9	50	50
K083	Ford Cortina 1600	2 of 40 DCOE 2	33	4.5	125	F16	155	50F9	40	55
K111	Jaguar XK-E	3 of 45 DCOE 13	38	3.5	165	F2	190	65F8	40	50
K124	Midget, Spitfire 1500	1 of 40 DCOE 18	28	4.5	125	F16	160	45F9	50	50
K125	Midget, Spitfire 1500	2 of 40 DCOE 18	30	4.5	115	F16	170	45F9	50	50
K126	Midget, Spitfire 1500	1 of 45 DCOE 13	34	3.5	150	F2	175	55F8	50	50
K131	Triumph TR-3, 4	2 of 40 DCOE 18	32	3.5	140	F15	150	50F2	50	50
K143	Triumph GT-6 '69-'73	3 of 40 DCOE 18	30	4.5	120	F2	160	50F11	45	Closed
K151 / K153	Triumph TR-6	3 of 40 DCOE 18	30	4.5	120	F11	160	50F11	40	50
K161	Triumph TR-7	2 of 40 DCOE 18	30	4.5	125	F9	155	55F8	45	45
K205	BMW 2002	1 ea. 45 DCOE 15 & 16	34	5.0	125	F9	180	45F8	40	70
K206	BMW 2002	2 of 40 DCOE 2	33	4.5	120	F16	190	50F8	40	55
K207 / K210	BMW Bavaria	2 of 32/36 DGAV / 2 of 32/36 DGAV-ICU	26/27	3.5/3.5	125/130	F50/F50	170/175	60/60	50	—
K208	BMW 6 cyl.	3 of 40 DCOE 18	30	4.5	120	F11	180	45F9	40	50
K216	Capri, Pinto 2000	1 of 45 DCOE 13	32	3.5	140	F2	180	55F2	50	50
K218	Capri, Pinto 2000	2 of 40 DCOE 2	32	4.5	125	F16	180	50F9	40	50
K219	Chevrolet Vega	1 of 40 DCOE 9	32	4.5	130	F16	170	55F8	60	40
K241	M-B 190SL	2 of 40 DCOE 18	30	4.5	125	F11	200	50F9	45	Closed
K288	Porsche 914 (2.0)	1 ea. 40 IDF 48 & 49	32	4.5	125	F16	180	50	50	55
K289	Porsche 924	1 ea. 45 DCOE 15 & 16								
K290 / K291	Porsche 356, 912	1 ea. 40 IDF 48 & 49	28	4.5	115	F11	200	50	50	55
K294	Porsche 924	2 of 40 DCOE 18	32	4.5	120	F5	180	45F9	50	55
K295	Porsche 914 (2.0)	1 ea. 44 IDF 50 & 51	36	4.5	135	F11	175	50	50	55
K296	Porsche 914 (2.0)	1 ea. 48 IDF 4 & 5	40	4.5	150	F2	180	55	50	Closed
K305	VW Dual Port Type 1	1 of 40 DCNF 12	28	4.5	130	F24	200	50	50	—
K317	VW Dual Port Type 1	1 ea. 40 IDF 48 & 49	28	4.5	115	F11	200	50	50	55
K319	VW Dual Port (1800+)	1 ea. 44 IDF 50 & 51	36	4.5	135	F11	175	50	50	55
K326	VW Dual Port (2000+)	1 ea. 48 IDF 4 & 5	40	4.5	150	F2	180	55	50	Closed
K327 / K328	VW Dual Port (2000+O)	2 of 48 IDA 4	37	4.5	145	F7	160	70F10	50	50
K347	Porsche/VW Type 4	1 ea. 40 IDF 48 & 49	28	4.5	120	F11	200	50	50	55
K348	Porsche/VW (1800+)	1 ea. 44 IDF 50 & 51	36	4.5	135	F11	175	50	50	55
K349	Porsche 914 (2.0)	1 ea. 48 IDF 4 & 5	40	4.5	150	F2	180	55	50	Closed
K403 / K454	VW Rabbit, Scirocco / Ford Escort	1 of 34 DMTR 21	25/26	4.0/4.0	125/120	F30F30	160/200	60/50	45	—

Conversion Kit Number	Application	Carburetor Quantity & Model	Main Venturi	Aux. Venturi	Main Jet	Em. Tube	Air Corr. Jet	Idle Jet	Acc. Pump Jet	Pump Ex. Valve
K404	VW Rabbit, Scirocco	1 ea. 40 IDF 13 & 15	28	4.5	120	F11	170	50	50	55
K407	VW Rabbit, Scirocco	2 of 40 DCOE 18	30	4.5	135	F9	220	45F9	40	50
K408	VW Rabbit, Scirocco	1 of 32 DFTA	22/22	4.0/4.0	100/110	F30/F30	170/200	50/50	45	—
K430	VW Rabbit, Scirocco	1 of 40 DCOE 18	29	4.5	130	F11	165	50F9	50	50
K458	Ford Escort	1 ea. 40 IDF 13 & 15	32	4.5	125	F11	210	55	40	80
K459	Ford Escort	2 of 40 DCOE 18								
K460	Ford Escort, Lynx	1 of 34 DATRA	23/27	4.0/4.0	112/130	F21/F25	210/180	50/50	45	—
K470	Pontiac Fiero	2 of 45 DCOE 9	36	4.5	130	F11	165	55F8	45	40
K480	Chrysler K2.2	2 of 40 DCOE 18	33	4.5	125	F11	175	45F9	40	40
K510	Fiat X1/9, 128	2 of 40 DCNF 12	32	4.5	130	F24	200	50	45	—
K511	Fiat X1/9, 128	2 of 40 DCOE 18	30	4.5						
K512	Fiat, 124, 131	1 ea. 40 IDF 13 & 15	32	4.5	125	F11	210	55	40	80
K514	Fiat 124, 131	2 of 40 DCOE 18	33	4.5	145	F16	220	50F9	35	55
K606	Chrysler 1600	1 of 40 DCOE 2	32	4.5	125	F20	160	50F8	50	50
K607	Chrysler 1600	2 of 40 DCOE 18	30	4.5	120	F16	185	45F9	40	50
K623	Datsun 1200	1 of 40 DCOE 18	27	4.5	110	F7	165	50F9	50	50
K625	Datsun 210	1 of 40 DCOE 2	28	4.5	120	F11	165	50F8	50	Closed
K635 K665	Datsun 610, 620, 710	1 of 40 DCOE 18	30	4.5	140	F11	175	45F9	50	55
K636 K666	Datsun 610, 710	1 of 45 DCOE 13								
K637, K667 K647, K669	Datsun 610, 710	2 of 40 DCOE 2	32	4.5	135	F15	170	55F2	50	50
K638	Datsun NAPS-Z	2 of 40 DCOE 2	33	4.5	120	F2	130	60F2	45	50
K659	Datsun 24 oz, 26 oz	3 of 40 DCOE 18	28	4.5	120	F11	170	50F9	40	55
K701	Honda Civic	1 of 40 DCOE 2	30	4.5	135	F16	180	45F9	35	35
K753	Toyota Corolla	1 of 40 DCOE 18	27	4.5	105	F7	155	50F11	50	50
K755	Toyota Corolla	1 of 40 DCOE 18	28	4.5	110	F15	200	45F9	50	50
K757	Toyota Corolla 2TC	2 of 40 DCOE 18	27	4.5	115	F16	175	45F9	45	50
K765	Toyota Corona	1 of 40 DCOE 18	30	4.5	135	F2	165	45F9	60	Closed
K767	Toyota Corona	2 of 40 DCOE 18	30	4.5	130	F2	170	45F9	45	Closed
K770 K777	Toyota 20R, 22R	2 of 40 DCOE 18	32	4.5	120	F11	190	50F8	40	40
K771 K776	Toyota 20R, 22R	2 of 45 DCOE 9	34	3.5	130	F2	180	50F9	45	50
K775	Toyota Celica 20R	1 of 45 DCOE 9	34	4.5	145	F15	150	50F8	60	Closed
K901	Volvo	2 of 40 DCOE 2	32	4.5	120	F11	190	50F8	40	40
K903 K904	Volvo	1 of 32/36 DGV	26/27	3.5/3.5	140/125	F50F6	185/160	60/50	50	—

©1984 Interco Parts Corp.

Weber Carburetor & Conversion Kit Suppliers

Alfa Ricambi
6644 San Fernando Road
Glendale, CA 91201
(818) 956-7933
(800) 221-3759
Alfa Romeo kits.

Autotech Sport Tuning
1800 N. Glassell St.
Orange, CA 92665
(714) 974-4600
Water-cooled VW specialists.

Bayless Racing, Inc.
1377 Barclay Circle
Marietta, GA 30060
(404) 422-6274
Fiat, Lancia kits.

CARTECH
11212 Goodnight Lane
Dallas, TX 93117
(214) 620-0389
Mazda, BMW, Toyota,
turbocharging specialists.

Centerline Products
4715 North Broadway
Boulder, CO 80302
(303) 447-0239
Alfa Romeo kits.

Colortune
White Eagle Manufacturing
RT1, Box 279
East Bernstadt, KY 40729
(606) 843-6126
Air/fuel mixture analyzer;
Weber tuning aid.

Fast Freddy's
2877 W. Lincoln
Anaheim, CA 92801
(714) 527-1812
General Weber parts and kits.

FAZA
Box 441
Montauk Hwy.
Westhampton, NY 11977
(516) 728-1992
Fiat, Lancia, Alfa Romeo kits.

Fuelish Parts, Ltd.
P.O. Box 152
Hollywood, FL 33020
(305) 922-0482
V8 specialists.

Greenfield Imported Car Parts
335 High Street
Greenfield, MA 01301
(413) 774-2819
General Weber parts and kits.

Inglese Induction Systems
P.O. Box 709
Branford, CT 06405
(203) 481-5544
V8 specialists.

INTERCO
150 Wireless Blvd.
Hauppauge, NY 11788
(516) 434-1818
(800) 645-7488
Weber East coast importer.
Extensive kit selection.

International Auto Parts
1309 Mountain View St.
Charlottesville, VA 22901
(804) 295-0127
Fiat, Lancia, Alfa Romeo kits.

JAM Engineering
886 Abrego St.
Monterey, CA 93942
(408) 372-1787 (CA)
(800) 431-3533 (Outside CA)
BMW & Mercedes Benz kits.

Korman Autoworks
1316 Headquarters Drive
Greensboro, NC 27405
(919) 272-9604
BMW kits.

PRO
135 17th Street
Santa Monica, CA 90402
(213) 393-5423
Porsche 911 specialists.

Red Lion
2895 W. 190th St.
Redondo Beach, CA 90278
(213) 376-0247 (CA)
(800) 722-8678 (Outside CA)
Air-cooled VW specialists.

Redline, Inc.
303 W. Artesia Blvd.
Compton, CA 90224
(213) 637-7774
(800) 421-2777
Weber West coast importer.
Extensive kit selection.

Shankle Automotive Engineering
9135-F Alabama Ave.
Chatsworth, CA 91311
(818) 709-6155
Alfa Romeo kits.

TMW Induction
400 Rutherford St.
Goleta, CA 93117
(805) 963-3329
General Weber parts.
Extensive kit selection.

World Distributors
511 Cypress Avenue
Venice, FL 34292
(800) 237-9270 (FL)
(800) 282-8538 (outside FL)
General Weber parts and kits.

Weber Work-alike Suppliers

CB Performance
28813 Farmersville Blvd.
Farmersville, CA 93223
(209) 733-8222
(800) 252-8337
Dellorto carburetor specialists.

Mikuni American Corp.
8910 Mikuni Ave.
Northridge, CA 91324
(213) 885-1242
Primary Mikuni distributor.

Nissan Motorsports
P.O. Box 191
Gardena, CA 90247
(213) 538-2610
Mikuni carburetors for racing Nissans.

TMW Induction
400 Rutherford St.
Goleta, CA 93117
(805) 963-3329
SK Racing Carburetor distributor.

METRIC/CUSTOMARY-UNIT EQUIVALENTS

Multiply:		by:		to get:	Multiply:		by:		to get:
LINEAR									
inches	X	25.4	=	millimeters(mm)		X	0.03937	=	inches
feet	X	0.3048	=	meters (m)		X	3.281	=	feet
miles	X	1.6093	=	kilometers (km)		X	0.6214	=	miles
AREA									
inches^2	X	645.16	=	millimeters^2(mm^2)		X	0.00155	=	inches^2
feet^2	X	0.0929	=	meters^2(m^2)		X	10.764	=	feet^2
VOLUME									
inches^3	X	16387	=	millimeters^3(mm^3)		X	0.000061	=	inches^3
inches^3	X	0.01639	=	liters (l)		X	61.024	=	inches^3
quarts	X	0.94635	=	liters (l)		X	1.0567	=	quarts
gallons	X	3.7854	=	liters (l)		X	0.2642	=	gallons
feet^3	X	28.317	=	liters (l)		X	0.03531	=	feet^3
feet^3	X	0.02832	=	meters^3(m^3)		X	35.315	=	feet^3
MASS									
pounds (av)	X	0.4536	=	kilograms (kg)		X	2.2046	=	pounds (av)
FORCE									
pounds—f(av)	X	4.448	=	newtons (N)		X	0.2248	=	pounds—f(av)
kilograms—f	X	9.807	=	newtons (N)		X	0.10197	=	kilograms—f

TEMPERATURE

Degrees Celsius (C) = 0.556 (F - 32)   Degrees Farenheit (F) = (1.8C) + 32

°F -40   0   32 40 80 98.6 120 160 200 212 240 280 320 °F
°C -40 -20   0   20   40   60   80 100 120 140 160   °C

Multiply:	by:	to get:	Multiply:	by:	to get:
ACCELERATION					
feet/sec^2	X 0.3048	= meters/sec^2(m/s^2)	X 3.281	= feet/sec^2	
inches/sec^2	X 0.0254	= meters/sec^2(m/s^2)	X 39.37	= inches/sec^2	
ENERGY OR WORK (Watt-second = joule = newton-meter)					
foot-pounds	X 1.3558	= joules (J)	X 0.7376	= foot-pounds	
calories	X 4.187	= joules (J)	X 0.2388	= calories	
Btu	X 1055	= joules (J)	X 0.000948	= Btu	
watt-hours	X 3600	= joules (J)	X 0.0002778	= watt-hours	
kilowatt-hrs	X 3.600	= megajoules (MJ)	X 0.2778	= kilowatt-hrs	
FUEL ECONOMY & FUEL CONSUMPTION					
miles/gal	X 0.42514	= kilometers/liter(km/l)	X 2.3522	= miles/gal	

Note:
235.2/(mi/gal) = liters/100km
235.2/(liters/100km) = mi/gal

PRESSURE OR STRESS					
inches Hg (60F)	X 3.377	= kilopascals (kPa)	X 0.2961	= inches Hg	
pounds/sq in.	X 6.895	= kilopascals (kPa)	X 0.145	= pounds/sq in	
POWER					
horsepower	X 0.746	= kilowatts (kW)	X 1.34	= horsepower	
TORQUE					
pound-inches	X 0.11298	= newton-meters (N-m)	X 8.851	= pound-inches	
pound-feet	X 1.3558	= newton-meters (N-m)	X 0.7376	= pound-feet	
pound-inches	X 0.0115	= kilogram-meters (Kg-M)	X 87	= pound-inches	
pound-feet	X 0.138	= kilogram-meters (Kg-M)	X 7.25	= pound-feet	
VELOCITY					
miles/hour	X 1.6093	= kilometers/hour(km/h)	X 0.6214	= miles/hour	
kilometers/hr	X 0.27778	= meters/sec (m/s)	X 3.600	= kilometers/hr	

COMMON METRIC PREFIXES

mega	(M)	= 1,000,000	or	10^6	centi	(c)	= 0.01	or	$10^{\times 2}$
kilo	(k)	= 1,000	or	10^3	milli	(m)	= 0.001	or	$10^{\times 3}$
hecto	(h)	= 100	or	10^2	micro	(µ)	= 0.000,001	or	$10^{\times 6}$

INDEX

Accelerator pump, 11, 25, 42
Air injection, 28
Air pump, 28
Air-correction jet, 12, 34, 136
Air/fuel (A/F) ratio, 5
Alf Francis, 8
Alfa Romeo, 12
Altitude Compensation Table, 166
Automatic intake valve, 16
Auxiliary circuits, 24
Auxiliary venturi choke, 41
Auxiliary venturi, 24
Brake Mean Effective Pressure
    (BMEP), 15, 115
Butterfly choke, 41
Butterfly valve, 25
Camshaft, 122
    Base circle, 122
    Duration, 122, 123
    Lift, 122, 123
    Ramp, 123
Carburetor icing, 25
Carburetor operation, 14
Carburetor selection, 126
Catalytic converter, 28
Charcoal canister, 27, 141
Choke plate, 25
Choke, 25, 27, 41
    Plunger valves, 41
Clearance volume, 15
Closed-loop control, 5
Colortune, 152
Combustion, 16
Compression ratio, 15, 116
Constant-Velocity Carburetors, 23
Conversion Kits, 168
DCOE Float Level Settings, 99
DCOE Initial Setup & Tuning, 99
DCOE Repair 84-100
    Accelerator pump, 88
    Idle circuit, 86
    Idle mixture screw, 91
    Main circuit, 87
    Progression holes, 87
    Split pin, 94
    Starter circuit, 87, 92
    Throttle shaft, 96
    Venturis, 95
DCOE and IDA Idle Jet Selection, 162
DCOE choke, 41
DFT Initial Setup & Tuning, 84

32/34 DFT Repair, 70-84
    Automatic choke, 83
    Dieseling, 71
    Electrically heated choke, 71
    Float-chamber vent valve, 71
    Fuel cut-off solenoid, 71, 76, 82
    Fuel-return check valve, 71
    Power valve, 71, 76
    Sealed idle mixture screw, 71
DGAV repair, 56-70
    Automatic choke, 62, 69
    Power valve, 59
DGAV Initial Setup & Tuning, 70
Dellorto, 9, 171
Detonation, 116
Diaphragm-type accelerator pump, 25
Displacement, 123
Distributor, 156
Diverter valve, 138, 141
Doppio Corpo Orrizontale, 86
Doppio Corpo, 11
Downdraft, 11
Duty-cycle solenoid, 5
Edoardo Weber, 6
Electronic Control Unit (ECU), 5
Emission control devices, 26
Emulsified fuel, 24
Emulsion tube selection, 163
Emulsion tube designation, 12
Emulsion tube, 6, 34, 37, 38
    Air volume, 38
    Hole orientation, 38
Emulsion tubes, 37
    Brake, 37
    Diameter, 37
    Step, 37
    Well, 37
End gas, 116
Equivalent Ratios, 5
Exhaust Gas Recirculation (EGR), 27, 28
Exploded View, 32 IMPE, 47
    Weber 40 DCNF, 7
    32 DFTA, 72
    32/36 DGAV, 56
    40/42/45 DCOE, 85
Fault identification, 32
Ferrari, 8, 11
Filter screen, 33
Flame front, 116
Flat spots, 25
Float chamber, 21

Float level adjustment, 136, 154
Float level, 154
Float-chamber vent valve, 27
Flow bench, 119
Four-stroke cycle, 16
Fuel injection, 143
Fuel bowl, 21
Fuel cut-off solenoid, 27
Fuel filter, 33
Fuel inlet nipple, 142
Fuel inlet valve, 21
Fuel level, 42
Fuel pressure regulator, 22, 154
Fuel puddles, 127
Fuel pump, 33
Fuel regulator, 152
Fuel-return check valve, 28
Fuel/air (F/A) ratio, 5
Gas, 17
Giovanni Battista Venturi, 21
Gulp valve, 138
Headers, 115
Hitachi/SU, 23
Horsepower, 17
IDA3C Initial Setup & Tuning, 114
IDA3C Repair, 100-114
    Air compensating adjustment, 100
    Air compensating screws, 106, 111
    Auxiliary venturis, 105
    Exploded view IDA, 102
    Float level, 113
    Nylock nuts, 103
    Progression circuit, 100
    Synchrometer, 112
    Synchronizing throttle valves, 110
    Unisyn, 112
IMPE Initial Setup & Tuning, 55
IMPE Repair, 45-55
    Cleaning, 51
    Float bowl, 48
    Needle valve, 49
    Throttle shaft, 53
    Throttle valve, 54
INTERCO, 9
Idle circuit, 22, 40
Idle jet, 12, 39, 136
Idle mixture adjustment screw, 24, 39
Idle-Circuit Emulsion Calibration, 12
Idle-speed screws, 135
Ignition timing, 117
Ignition timing advance, 117

INDEX

Ignition timing advance, 117
Ignition timing retard, 117
Ignition timing, 156
Ignition, 33
Inertia, 17, 19, 20
Inglese Induction Systems, 153
Inglese Troubleshooting Guide, 159
Injector nozzles, 28
Invertito, 11
JAM Engineering, 8
Jet size designation, 12
Jetting, 136
Knock, 116
Lamborghini, 11
Main circuit, 24, 34
Main jet, 12, 34, 37, 136
Main venturi diameter, 35, 36
Main venturi types, 35
Main venturi, 24, 35
Main-jet locations, 36
Manifold design, 127
Manifold heater, 141
Manifold selection, 126
Manifold types, 131, 133
    Cross-H, 131
    Dual-plane, 131
    Isolated-runner, 131, 133
    Plenum-ram, 131
    Single-plane, 131
Maserati, 8
Mass, 17
Metric conversion, 35
Metric/Customary-Unit Equivalents, 172
Mikuni, 9
Mixture dynamics, 18
Mixture flow, 16
Momentum, 17, 19, 121
NASCAR, 30
Needle valve, 21
Octane rating, 116
Orizzontale, 11
Otto Cycle, 16
    Compression, 16
    Exhaust, 16
    Intake, 16
    Power, 16
Overlap, 19, 122
Oxidation converters, 28
PCV, 29
Piston-type accelerator pump, 25
Plunger valves, 41

Poppet valves, 18
Porsche, 12
Port openings, 121
Porting, 121
Power Valves Values, 26
Power valve, 26, 42
Pressure differance, 16
Pressure waves, 120
Production Jetting Table, 160
Progression Circuit, 24, 39
Progression holes, 24
Ram tuning, 120, 121
Reciprocating motion, 15
Redline, 137
Reduction converters, 28
Repair methods, 42
SPICA, 143
Scavenging effect, 18
Series number, 126
Shankle Engineering, 143
Signal, 20, 38
Soft mounts, 150
Solenoid-controlled valve, 5
Sparkplugs, 33
Sparkplugs, reading, 34
Standard Jetting, 167
Standard Tunings, 161
Starter circuit, 25
Stoichiometric, 5, 20
"Strangler," 25
Stroke, 124
Supercharging/Turbocharging, 125
Suppliers List, 170
Surface carburetor, 16
Synchrometer, 135
Synchronization, 151
Synchronizing multiple carburetors, 135-136, 156, 158
TWM Induction, 9, 37
Tables & Charts
    DCOE Float Level Settings, 99
    Equivalent Ratio, 5
    Idle-Circuit Emulsion Calibration, 12
    Main venturi diameter, 35, 36
    Power Valves Values, 26
    Weber Carburetor Types, 13
Thermostatic actuator, 144
Throttle linkage geometry, 133
Throttle plate locations, 40
Throttle plate, 40
Throttle valve, 54

Timing tape, 156
Troubleshooting, 30
    Deduction, 31
    Logic, 31
    Observation, 31
Turbocharging, 125
Unisyn, 135
Vacuum pull-off, 27
Vacuum, 17
Vapor lock, 28
Vent nipple, 142
Venturi, 6, 20, 23
Volumetric efficiency (VE), 18, 119
Wavefront, 121
Weber Carburetor Orientation Diagrams, 128-130
Weber Carburetor Types, 13
Weber Gasket Kits and Repair Kits, 166
Weber Installation and Mounting Parts, 165
Weber Special Service Tools, 164
Weber conversions, 137
Weber diagnosis, 32
Weber history, 6
Weber model numbers, 11
Weber mounting diagrams, 132
Weber selection, 134
    Auxiliary venturi, 135
    Main venturis, 134
    Secondary venturi, 135

Tune Up
with HP Automotive Handbooks

__ **1,001 High Tech Performance Tips** by Wayne Scraba 1-55788-199-5/$16.00
__ **Auto Math Handbook** by John Lawlor 1-55788-020-4/$15.95
__ **Automotive Dictionary** by John Edwards 1-55788-067-0/$24.95
__ **Automotive Electrical Handbook** by Jim Horner 0-89586-238-7/$15.95
__ **Automotive Paint Handbook** by Jim Pfanstiehl 1-55788-034-4/$15.95
__ **Brake Handbook** by Fred Puhn 0-89586-232-8/$14.95
__ **Camaro Performance Handbook** by David Shelby 1-55788-057-3/$16.95
__ **Camaro Restoration Handbook** by Tom Currao and Ron Sessions 0-89586-375-8/$16.00
__ **Car Collector's Handbook** by Peter Sessler 1-55788-039-5/$14.95
__ **Chevrolet Power** edited by Rich Voegelin 1-55788-087-5/$19.95
__ **Classic Car Restorer's Handbook** by Jim Richardson 1-55788-194-4/$15.00
__ **Holley Carburetors, Manifolds and Fuel Injection (Revised Edition)** by Bill Fisher and Mike Urich 1-55788-052-2/$17.00
__ **How to Make Your Car Handle** by Fred Puhn 0-912-65689-1/$16.00
__ **Metal Fabricator's Handbook** by Ron Fournier 0-89586-870-9/$15.95
__ **Mustang Performance Handbook** by William R. Mathis 1-55788-193-6/$15.95
__ **Mustang Performance Handbook 2** by William R. Mathis 1-55788-202-9/$15.00
__ **Mustang Restoration Handbook** by Don Taylor 0-89586-402-9/$15.95
__ **Mustang Weekend Projects 1964 ½—1967** by Jerry Heasley 1-55788-230-4/$17.00
__ **Paint & Body Handbook (Revised Edition)** by Don Taylor and Larry Hofer 1-55788-082-4/$15.00
__ **Race Car Engineering & Mechanics** by Paul Van Valkenburgh 1-55788-064-6/$16.95
__ **Sheet Metal Handbook** by Ron and Sue Fournier 0-89586-757-5/$15.95
__ **Street Rodder's Handbook** by Frank Oddo 0-89586-369-3/$16.00
__ **Turbo Hydramatic 350** by Ron Sessions 0-89586-051-1/$15.95
__ **Turbochargers** by Hugh MacInnes 0-89586-135-6/$15.95
__ **Understanding Automotive Emissions Control** by Larry Carley and Bob Freudenberger 1-55788-201-0/$16.00
__ **Welder's Handbook** by Richard Finch 0-89586-257-3/$14.95

Payable in U.S. funds. No cash orders accepted. Postage & handling: $1.75 for one book, 75¢ for each additional. Maximum postage $5.50. Prices, postage and handling charges may change without notice. Visa, Amex, MasterCard call 1-800-788-6262, ext. 1, refer to ad #583

Or check above books and send this order form to:
The Berkley Publishing Group
390 Murray Hill Pkwy., Dept. B
East Rutherford, NJ 07073

Bill my: ☐ Visa ☐ MasterCard ☐ Amex expires _____

Card #_____
($15 minimum)
Signature_____

Please allow 6 weeks for delivery

Name_____

Address_____

City_____

State/ZIP_____

Or enclosed is my: ☐ check ☐ money order

Book Total $_____
Postage & Handling $_____
Applicable Sales Tax $_____
(NY, NJ, PA, CA, GST Can.)
Total Amount Due $_____